Timeless

EVOLUTION of TRADE and MONEY

*Towards
Sustainable Economies*

John Wnuk

ISBN: 978-0-578-68517-5

Evolution of trade and money is timeless.

Trade has become the peaceful alternative to war. It is the sharing of abundance over worldwide trade routes--the modern silk roads.

Money simplified trade compared to barter, and allows more choices in life. It is the fuel for lifestyle.

The Internet began a new paradigm for digital transformation in trade, money, and economies.

Now there are more choices--to live in the past or diversify.

Contents

Contents

Timeless

A. END GAME

The evolution of trade and money is timeless--and includes many mysteries. The end game of this book is to provide a view of that evolution from outside the financial box. A view from outside the box can remove the mysteries and reveal possible opportunities. That view is also about technology, with the Internet having a significant impact. It is also about innovations, habits, and values--that can provide clues for what may be next. *(Ref's. A1, A2)*

> *Ref. A1:*
> *investopedia.com/articles/07/roots_of_money.asp*
> *Ref. A2:*
> *investopedia.com/.../how-the-internet-has-changed-investing.aspx*

1) Innovations. Humankind tends to favor innovations that offer time and cost savings or lifestyle benefits. Examples are electric lights vs. candles, telephone vs. telegraph, and working from home using a computer and Internet vs. travel and working at an office location. Advances in computers, communications and Augmented Intelligence (AI), are modern tools for pioneers of innovations of tomorrow. *(Ref. A3)*

> *Ref. A3:*
> *whatis.techtarget.com/definition/augmented-intelligence*

2) Habits. Money habits can challenge reason. There is a bond to the U.S. Dollar (USD) that is not backed by anything tangible. Many still think the dollar is backed by gold, but it is not. The dollar went off the gold standard in 1971, and the dollar as paper or deposit does not have intrinsic value. *(Ref. A4)*

Also, the term "fiat" can be misleading. For many, it is a car, but the dollar is fiat money. That means it has value by decree, backed only by debt. *(Ref. A5)*

> *Ref. A4:*
> *www.federalreservehistory.org/essays/gold_convertibility_ends*
> *Ref. A5:*
> *usdebtclock.org/*

3) Values. The fiat dollar does have value as a convenient means of trade. The dollar also replaced the British pound as the world reserve currency in 1944, as other countries were confident that the U.S. could pay its debts. This bond to fiat money, however, can hinder an understanding of what is new. *(Ref. A6)*

For example, in 2009 an innovation based on mathematics and cryptography was created. That innovation was an Electronic Cash System called "Bitcoin," and that system included a "Blockchain" worldwide, public ledger of peer-to-peer transactions. *(Ref. A7)*

Bitcoin had no established value until May 22nd, 2010. At that time, a computer programmer, Laszlo Hanyecz, wanted to know if he could buy anything with his Bitcoin mining rewards. He ordered two large pizzas worth about $30 at the time, and offered to pay with 10,000 Bitcoins. Laszlo enjoyed the pizzas with his family, and that event is now celebrated every year on May 22nd as Bitcoin Pizza Day. *(Ref. A8)*

In May 2010, a Bitcoin (BTC) was worth roughly three for a penny (i.e., about $30/10,000 BTC = $0.003/BTC), and in September 2020, that 10,000 BTC was worth over $100 million. Wow.

With cryptography as a base, Bitcoin is also referred to as a "cryptocurrency," or "crypto." Other popular terms include "virtual currency," "Internet money," "modern money," "new money," "digital asset," "crypto asset," and "coin."

With the success of Bitcoin, other crypto or "Altcoins" soon followed each with their own blockchain or Distributed Ledger Technology (DLT). The mechanics of fiat and crypto systems, however, continues to be a mystery to most. This book is intended to remove that mystery, so that more can enjoy the opportunity to diversify with modern methods for trade and money. *(Ref. A9)*

Ref. A6:
investopedia.com/...how-us-dollar-became-worlds-reserve-currency.asp
Ref. A7:
bitcoin.org/bitcoin.pdf

Ref. A8:
finance.yahoo.com/news/bitcoin-pizza-day-celebrating-world...html
Ref. A9:
coinmarketcap.com/all/views/all/

B. MYSTERIES

The Federal Reserve System (Fed) has existed for over a century, but why it exists and how the fiat money system works is a mystery to many. Understanding the Fed and fiat is an important step towards understanding crypto. Mystery also surrounds Satoshi Nakamoto and his Bitcoin Electronic Cash System. A lack of understanding of money basics can limit or misguide choices in life.

1) Fed. Over a century ago, on a so called "duck hunt" on Jekyll Island, several wealthy U.S. business persons met in secret and established the foundations for a national financial system or central bank for the U.S. That system, created to prevent bank failures, now sets monetary policy, prints and mints fiat money as needed, and generates billions in profit that is sent to the U.S. Treasury every year. However, is it a mystery that a central bank is reporting billions in profit every year for a nation that is trillions in debt? *(Ref's. B1, B2).*

Ref. B1:
federalreservehistory.org/essays/jekyll_island_conference
Ref. B2:
federalreserve.gov/newsevents/pressreleases/other20190110a.htm

2) Satoshi. The inventor of Bitcoin and Blockchain has been a mystery for over a decade. Who is Satoshi Nakamoto? There are lots of clues about one of the richest persons on earth, but no answers. One wonders why remaining anonymous was a chosen lifestyle? *(Ref. B3)*

Ref. B3:
en.bitcoin.it/wiki/Satoshi_Nakamoto

3) Electronic Cash. Another mystery is more complex. How could volatile Bitcoins achieve billions in market capitalization in a few years, while aging fiat with so called "stable" value has trillions in spiraling debt? Sadly, the fiat debt question tends to be avoided, or worse, put down by naysayers with distorted views suggesting that insidious spiraling debt is OK, but healthy market volatility is not. *(Ref. B4)*

Ref. B4:
coinmarketcap.com/currencies/bitcoin/

4) Choices. What is new can tend to be distrusted at first. Considering the evolution of trade and money, there is something new all the time--from barter to brokers to banks to blockchains, and to what is next. It is a choice to participate in that evolution. Many choose not to participate and prefer to live in the comfort of the past. Those that choose to participate have an opportunity to diversify. They can become more informed and connect the dots between the limited tangibles of the past and the unlimited future creations of humankind.

To help make better choices, suggested reading in this book are provided for **a)** Newbies, **b)** Wannabe Experts, **c)** Crypto Futurists, with something for all in **d)** Brief Digest.

a) Newbies. Start by learning about the mechanics of the dollar, and then compare with the mechanics of crypto. Start with Section C. FIAT MECHANICS on pages 7-9, and then read Section E. COMPARISON: CRYPTO vs. FIAT on pages 19-21.

b) Wannabe Experts. Learn more by experiencing the process and getting some skin in the game. Read the Epilogue pages 74-77. Get a Bitcoin wallet app on a smartphone. Then on a computer, get an account at Coinbase.com and read the pricing and fees. Link the Coinbase account to a bank or credit card--click on settings, then Payment methods, then enter bank or credit card data. Coinbase then checks that credentials are valid. Invest some "training" money, and buy 0.01 Bitcoin (BTC) on Coinbase. Then send say 0.005 BTC from Coinbase to the smartphone wallet. Observe that transaction on a Blockchain explorer and any fees. If you don't say Wow after that exercise, then maybe crypto is not for you. *(Ref's. B5, B6, B7)*

c) Crypto Futurists. Start by learning more about what is beyond the blockchain. Read Section F. BEYOND THE BLOCKCHAIN starting on page 22. Section F covers green distributed ledgers that are fast, fee-less, and miner-less. Continue with Section G. TOWARDS SUSTAINABLE ECONOMIES that starts on page 28, and then the Appendices. That material has many visionary concepts for income, credit, payments, voting, and more.

d) Brief Digest. Of the thousands of existing crypto, which could likely flourish in the future? No one knows for sure, but blockchain-based crypto will likely continue including Bitcoin (BTC) as digital gold, Ethereum (ETH) as a programmable platform for Altcoins, and Cardano (ADA) with a comprehensive long term development roadmap. Beyond the blockchain, an effective and efficient (i.e., green) criteria could identify crypto and DLT that will flourish. Efficient implies methods that are fast, economic, and secure. Effextive implies the right methods are used for transaction verification and ledger accounting. That green criteria could filter to at least IOTA (MIOTA), Nano (NANO) and Holo (HOT) as described on pages 22-23, and c) on page 45. *(Ref's. B8, B9, B10, B11, B12, B13).*

Ref. B5:
geckoandfly.com/23587/bitcoin-wallet-ios-android/
Ref. B6:
coinbase.com/
Ref. B7:
blockchain.com/explorer
Ref. B8:
coinmarketcap.com/currencies/ethereum/
Ref. B9:
coinmarketcap.com/currencies/cardano/
Ref. B10:
cardanoroadmap.com/en/shelley/
Ref. B11:
coinmarketcap.com/currencies/iota/
Ref. B12:
coinmarketcap.com/currencies/nano/
Ref. B13:
coinmarketcap.com/currencies/holo/

C. FIAT MECHANICS

In the beginning, barter was the method of trade with values determined by a consensus of peers. Over the millennia came third parties with values determined by brokers and banks. The method of trade also evolved from physical material, such as cattle and cowrie shells, to printed paper and minted coins with values initially backed by gold.

That all changed with fiat money backed by decree and debt. The best way to characterize fiat is by its mechanics including **a)** value, **b)** policy, **c)** seigniorage, **d)** fractional reserve banking, **e)** security, **f)** clearing, **g)** settlements, **h)** counterfeits, **i)** risk, and **j)** sovereignty.

a) Fiat value is not backed by anything tangible. It is backed by sovereign decree and debt in the trillions. *(Ref. C1)*

Ref. C1:
usdebtclock.org/world-debt-clock.html

b) Central banks (e.g., the Fed in the U.S.) control monetary policy including how much physical fiat money is to be printed and minted each year by a tax-funded treasury. *(Ref. C2)*

Ref. C2:
federalreserve.gov/monetarypolicy/2019-07-mpr-summary.htm

c) Central banks make profits several ways including seigniorage tax on fiat sold to banks. Seigniorage is the difference between cost to produce and the face value of the money. For example, in 2019 it cost the U.S. Treasury on average $0.142 to print a $100 bill that was sold to banks at face value (i.e., $100)--with a profit at over 70,000%. *(Ref. C3, C4)*

Ref. C3:
investopedia.com/terms/s/seigniorage.asp
Ref. C4:
federalreserve.gov/faqs/currency_12771.htm

d) Banks make profits through fractional reserve lending, meaning that most fiat deposits are not in the bank's vault. *(Ref. C5)*

Ref. C5:
investopedia.com/terms/f/fractionalreservebanking.asp

e) Fiat bank account security is through a login User Identification (ID), strong password, and optional 2-Factor Authentication (2FA). *(Ref. C6)*

Ref. C6:
investopedia.com/terms/t/twofactor-authentication-2fa.asp

f) Clearing of transactions between fiat bank accounts takes days through an Automated Clearing House (ACH). *(Ref. C7)*

Ref. C7:
investopedia.com/terms/a/ach.asp

g) Fiat settlements between banks use a Real Time Gross Settlements (RTGS) system, and there is an RTGS for each sovereign nation (e.g., Fedwire in the U.S.). *(Ref. C8)*

Ref. C8:
en.wikipedia.org/wiki/Real-time_gross_settlement

h) Fiat counterfeits are estimated by the U.S. Treasury to be $70 to $200 million in the U.S., implying possible billions in counterfeits worldwide. *(Ref. C9)*

Ref. C9:
en.wikipedia.org/wiki/Counterfeit_United_States_currency

i) Fiat is favored for deposits insured by Federal Deposit Insurance Corporation (FDIC), low risk investments, public and private legal financial transactions; and private, untraceable illegal financial transactions. *(Ref. C10)*

Ref. C10:
cryptoiq.co/...fiat-currency-is-used-for-crime-1000-2500-times-more-than-crypto/

j) Fiat alternatives include a variety of printed paper and minted coins controlled by each sovereign nation. There are Dollars in the U.S, Euros in Europe, Rubles in Russia, and so on, along with frequently changing fiat exchange rates. *(Ref. C11)*

Ref. C11:
travelex.com/currency/current-world-currencies

D. CRYPTO MECHANICS

Before making a comparison with fiat, more knowledge on crypto mechanics would be helpful. Crypto mechanics would include more on wallets, transactions, mining, money creation, making change, volatility, regulation, smart contracts, and recourse for problems.

1) Wallets. Money needs methods for production, storage, transfer, and security. For fiat, there are treasury, banks, and physical wallets with cash, credit cards, ID, and more. For crypto, there are digital wallets. A digital wallet is a paired public address and private key. Also paired with the private key is a public key to verify a digital signature for a secure transaction over an insecure Internet. The basis for Bitcoin private key, public key, and public address pairing is mathematics as defined in the Elliptic Curve Digital Signature Algorithm (ECDSA). *(Ref. D1)*

The digital wallet's public address and private key have associated Quick Response (QR) codes for computer and smartphone readability. The public address is like a public mailbox address to receive Bitcoins from anyone. The private key is like a mailbox key that only the owner can use to withdraw Bitcoins. A Bitcoin app on a computer or smartphone can withdraw Bitcoins from a wallet by optically sweeping the QR code for the private key. If the private key is managed by a third party such as Coinbase or Binance, that third party is the owner of the value and operates similar to a bank with a promise to pay on demand. *(Ref. D2)*

Ref. D1:
en.bitcoin.it/wiki/Elliptic_Curve_Digital_Signature_Algorithm
Ref. D2:
en.wikipedia.org/wiki/QR_code

2) Transactions. Transactions between peer wallets are pseudo-anonymous, and use Public Key Cryptography (PKC) for security. Peer wallets are the same type such as Bitcoin to Bitcoin, vs. non-peer Bitcoin and Ethereum wallets. Pseudo-anonymous means the wallet transactions are traceable on a blockchain or DLT, but the owner of a wallet can be difficult to identify without legal subpoenas for private information.

PKC provides a method for secure communications over an insecure network. A signing algorithm produces a digital signature using the message and private key. A verification algorithm, with the digital signature, message and public key, identifies the sender as the owner of the private key and that message content (i.e., value being transferred) has not been altered, without revealing the private key to the receiver. *(Ref. D3)*

An invalid transaction between non-peer wallets, or with an error in content or format between peer wallets, can result in lost value if not detected and then rejected before being cleared for entry to a blockchain. This would be similar to a bad check getting rejected by an Automated Teller Machine (ATM) and not accepted as a fiat deposit.

For exchanges between non-peer wallets or between a crypto wallet and fiat bank account, an exchange facility such Coinbase could be used. For that exchange service, there is a fee to cover the costs of the exchange facility's operation. *(Ref. D4)*

Ref. D3:
en.bitcoin.it/wiki/How_bitcoin_works
Ref. D4:
investopedia.com/terms/a/atm.asp

3) Mining. Money follows technology evolution, with the Internet having a significant impact. Before the Internet, mining was about physical work with machines and picks and shovels to find tangible resources. After the Internet, the concept of "mining" expanded to include the work of computers.

Bitcoin first appeared in 2009 as an Internet protocol. That protocol can transfer value between peer wallets and includes a fee-based security service called "mining." Bitcoin mining uses computers to "clear" valid transactions before entry to the Blockchain. Sending Bitcoin to the wrong address can be an error on the part of the sender, possibly resulting in permanent loss of that value, if not otherwise rejected as an invalid transaction. *(Ref. D5)*

The Blockchain shows "cleared" transactions with no confirmations, and "settled" transactions with several confirmation. It takes about 10 minutes for the first confirmation of a valid block containing about 4,000 Bitcoin transactions. If a block error is detected by the mining process, there is additional processing for the unconfirmed block with added delay for transaction settlements.

Ref. D5:
investopedia.com/terms/b/bitcoin-mining.asp

4) Money Creation. How are new Bitcoins created? In the mining process described previously, miners (or groups of miners in mining pools) must also solve a mathematical challenge. The first to do so is rewarded with new Bitcoins plus mining fees for each confirmed block. The reward is deposited to the winner's wallet as the first transaction in a newly confirmed block.

In the beginning in 2009, that reward was 50 new Bitcoins about every 10 minutes. The reward interval is a Bitcoin protocol parameter, which can vary depending on the Altcoin. For example, Ethereum has a reward interval that varies from 10 to 20 seconds. *(Ref. D6)*

In 2020, the reward was 6.25 new Bitcoins about every 10 minutes. Cutting the reward in half is a Bitcoin protocol parameter referred to as "halving" that occurs every 210,000 blocks or about every four years. The last Bitcoin reward will be in 2140, with only mining fees as rewards to continue after 2140. The reward method for Bitcoin is referred to as Proof Of Work (POW). Halving intervals, if any, and alternative methods such as Proof Of Stake (POS), can vary with Altcoins. Ethereum does not have a pre-determined halving interval and is evolving to POS. *(Ref. D7)*

The Bitcoin reward reduction is modeled after physical mining. For example, gold mining tends to produce less in time due to the difficulty in finding a reduced supply, but that reduced supply of gold typically has more value. Bitcoin has a similar process with reduced supply and increased value in time. For example, in 2009 Bitcoin had no meaningful value (i.e., crypto dust). In 2010, one Bitcoin was worth less than a penny but has increased to thousands in value over time.

Crypto has mining alternatives. High energy computer facilities are used for mining Bitcoin and Ethereum. A low energy "pre-mined" alternative was used for Ripple (XRP). Pre-mined means there are no miners, and all the crypto is made available in the genesis block, or alternative method depending on the crypto. Mining results in a distributed process for miners that receive new crypto, and that new crypto is typically labeled as a "currency." Pre-mined crypto is more centralized in regards to who gets the rewards, and could be labeled as a "security" with possible regulation by the Securities and Exchange Commission (SEC). *(Ref's. D8, D9)*

Ref. D6:
en.bitcoin.it/wiki/Controlled_supply
Ref. D7:
bitcoinblockhalf.com/
Ref. D8:
investopedia.com/terms/p/premining.asp
Ref. D9:
investopedia.com/news/how-sec-regs-will-change-cryptocurrency-markets/

5) Making Change. A subtle concept is making change in a crypto transaction. It is similar to making change in a fiat transaction where a customer uses $10 to buy a $2 cup of coffee. The $10 is traded for the coffee and $8 in change. Consider a Bitcoin transaction where a source wallet1 has 2 BTC and is used to purchase an item that costs say 0.5 BTC to be deposited at destination wallet2. The 2 BTC are withdrawn from wallet1, 0.5 BTC are sent to destination wallet2 to purchase the item, and the protocol returns the change (i.e., balance 1.5 BTC less transaction fee) to a new source wallet3.

That Bitcoin transaction is recorded on the Blockchain as one source (wallet1) and two destinations (wallet2 and wallet3). There are variations in this protocol depending on the Altcoin. For example, with Ethereum (ETH) the source wallet1 would be reduced to a new balance, and there is no new destination wallet3 for change. The Ethereum process was a design choice for simplicity.

6) Volatility. Why so much? Crypto whales (e.g., early adopters and institutional investors) and the law of networks are likely causes. Early adopters were those with mining facilities and the first to be rewarded with new crypto, along with those that bought in for pennies. Institutional investors are recent adopters that manage large financial portfolios. Crypto whales can be responsible for "pump and dump" schemes. They can easily buy large quantities and pump up values, or dump (i.e., sell) and drive down crypto values.

For law of networks, effects are proportional to equation n(n-1)/2, where n is a function of number of users and number of use cases. This equation developed for phone networks such that with 1 phone the value would be 1x0/2 = 0, 2 phones 2x1/2 = 1, 3 phones 3x2/2 = 3, 4 phones 4x3/2 = 6, and so on with exponential increases with large n. *(Ref. D10)*

Bitcoin started with a few early adopters and now there are millions of users. There are risks but no discrimination on who participates, and it is a way for anybody to diversify with a high risk investment. For use cases, there is store of value, investments, day trading, payments, gifts, and exchanges between different types of crypto. For example, Bitcoin can be exchanged with thousands of other Altcoins, through hundreds of crypto exchanges in many countries.

Soon there will be more stablecoins, other than Tether (USDT) and USD Coin (USDC), with values pegged to fiat or a tangible asset. New stablecoins can come from sovereign nations, financial organizations, or social networks with billions of users. The law of networks with n(n-1)/2 can become exponentially large with a range of values from n(low) = users + use cases, to n(high) = users x use cases. *(Ref. D11)*

While there is a lower limit to Bitcoin value, there is no higher limit. Changes in Bitcoin's fiat value from less than $0.01 to near $20,000 are part of history. Who is to say that higher values of $100,000 or more will never occur? *(Ref. D12)*

Crypto volatility implies significant profit or loss potential. Profits come from buying low and selling high. For the uninformed trying to get rich quick, the result can be a risk of loss by buying high and selling low. For beginners, the best investment is not money but time to learn more about risks and benefits of diversification. *(Ref. D13)*

Ref. D10:
collaboration.fandom.com/wiki/Metcalfe%E2%80%99s_Law
Ref. D11:
investopedia.com/terms/s/stablecoin.asp
Ref. D12:
coinswitch.co/news/bitcoin-price-prediction-2020-2025-latest-btc-price-prediction-bitcoin-news-update
Ref. D13:
themuse.com/advice/the-7-best-ways-to-invest-your-time

7) Regulation. In the U.S., crypto is regulated more ways than fiat. For fiat, a money services business is required to have a Money Transmitter License (MTL) and comply with Know Your Client (KYC) and Anti-Money Laundering (AML) regulations. A money services business, as defined by Financial Crimes Enforcement Network (FinCEN), can include one or more of several services from currency exchanger to the U.S. Postal Service. *(Ref's. D14, D15, D16)*

Crypto, however, is regulated four different ways--as money, property, commodity, and security. Money regulation for crypto is the same as fiat, as specified by FinCEN. Property regulation is from the Internal Revenue Service (IRS), with guidance that taxes must be paid for crypto profit/loss the same as physical property profit/loss. Commodity regulation is handled by the Commodities Futures Trading Commission (CFTC) that claims a role in emerging crypto innovations. Security regulation was discussed previously where it is possible that pre-mined crypto could be labeled as a "security" and regulated by the SEC. *(Ref's. D17, D18)*

In time, there will likely be separate regulations for a crypto asset class similar to efforts in Japan. Japan has the Payment Services Act (PSA) that is scheduled to be effective in 2020. Crypto is used worldwide, with regulations depending on the country, and it is not permitted or illegal where freedoms are restricted. *(Ref. D19)*

Ref. D14:
licenselogix.com/faq/who-needs-a-money-transmitter-license
Ref. D15:
investopedia.com/terms/a/aml.asp
Ref. D16:
investopedia.com/terms/f/fincen.asp
Ref. D17:
irs.gov/businesses/small-businesses-self-employed/virtual-currencies
Ref. D18:
cftc.gov/Bitcoin/index.htm
Ref. D1:
https://www.loc.gov/law/...Payment...Services...Act...Japan...

8) Smart Contracts. Contracts are legally binding agreements between two or more parties. Typically, the contract is documented in a computer file or on paper and signed by affected parties. Consequence of breaking a contract between two private parties is not illegal but is considered to be a "breach" of contract. When there is a breach of contract, it is up to the party or parties that has been unfairly treated to pursue damages in a civil court of law. *(Ref. D20)*

Smart contracts are permanently recorded on a Blockchain or DLT, that cannot be changed, lost, or stolen. An example of a smart contract on an Ethereum Blockchain. An LLC has been created with two primary members, and it allows others to buy into the LLC. This smart contract could have public and private parts and details per this reference. *(Ref. D21)*

Timeless

A classic example of a smart contract is the U.S. Electoral System. It is a contract that arbitrates between "justice" and "truth" in an election. The electoral system can provide fairness (i.e., justice) by preventing a popular vote (i.e., truth) from a clustered majority of voters (i.e., over 50%) in a few cities from taking control of a nation with over 19,000 cities, towns, villages, and boroughs. *(Ref. D22)*

Ref. D20:
upcounsel.com/consequences-of-breaking-a-contract
Ref. D21:
jmwnuk.wixsite.com/digitalassets/p-template
Ref. D22:
fl-pda.org/independent/courses/elementary/socialScience/section3/3e.htm

9) Recourse. What to do if something goes wrong? With fiat banks, deposit accounts have FDIC insurance to protect against a bank's failure. Fiat merchant accounts have purchase, return, repair or replace policies, warranties, loss protection, and a customer service phone number. With crypto, the focus is on prevention. Opening a crypto account may seem more complex, but the intent is to prevent the need for recourse. Prevention tools include **a)** registration, **b)** authentication, **c)** accuracy, **d)** delay, **e)** lockout, and **f)** help desk.

a) Registration for a new account on a crypto exchange begins with a user id, a strong password, and KYC information to determine the allowed monetary transactions. The KYC information can include an email address, photo of a driver's license or passport, and/or a copy of a utility bill.

b) Authentication at login begins with a user id and password, plus optional 2FA with entry of a time limited security code from a second source. A test to solve a puzzle, with pictures or patterns, could also be part of authentication to verify that the login is from a person and not a hack attempt from a computer. If the device or computer operating system changes from the original registration, the authentication process could be re-initiated and require 2FA. If the users location changes from the original registration, the Internet Service Provider (ISP) could also change and the authentication process could be re-initiated

c) The accuracy of actions can be checked for format and content with a required response to correct errors or approve the transaction.

d) A delay can occur after all actions are completed, to allow time to make further changes or to cancel the transaction.

e) A lockout can occur if there are too many login errors, or the authentication process must be re-initiated too many times.

f) The help desk can be contacted for next steps if a registered user gets locked out, with unregistered users or hackers getting blocked from entry.

E. COMPARISON: CRYPTO vs. FIAT

Crypto and fiat mechanics have significant differences. This section compares: **a)** values, **b)** costs, **c)** seigniorage, **d)** fractional reserve banking, **e)** security, **f)** clearing, **g)** settlements, **h)** counterfeits, **i)** diversification, and **j)** location of use.

a) Crypto value, not including stablecoins, is determined by a world marketplace vs. fiat value that is determined by sovereign decree.

b) Crypto's digital production and distribution has time and cost savings vs. fiat's tax-funded physical production, storage, transport, and security for printed paper and minted coins.

c) Crypto can have transaction fees, but no hidden seigniorage tax as in the huge markups in fiat money sold to banks.

d) Crypto digital wallets store all the value while fiat banks, with fractional reserve banking, do not store all the deposits in the bank's vault.

e) Crypto exchange login security is similar to fiat bank accounts with user ID's, strong passwords, and 2FA, but digital wallets have the additional cryptographic security of private keys that can be passphrase protected.

f) Miners clear crypto transactions for entry to the blockchain in minutes vs. fiat's ACH that takes days.

g) There is a separate blockchain or DLT for worldwide settlements for each type of crypto. Fiat has separate and costly RTGS for each sovereign nation. And note that the blockchain/DLT and RTGS provide common one-way, irreversible transactions.

h) Crypto is designed to prevent double spends. That would be similar to a capability (that does not exist) to prevent spending of counterfeit fiat money--with possibly billions of fiat counterfeits worldwide.

i) Crypto is favored for diversification for high-risk financial investments or low risk USDC that pays interest (e.g., at Coinbase in 2020), but is **not** favored for private, untraceable illegal financial transactions as in fiat, since wallet transactions are traceable on the blockchain/DLT. *(Ref. E1)*

Ref. E1:
cryptoiq.co/...fiat-currency-is-used-for-crime-1000-2500-times-more-than-crypto/

j) Crypto can be used worldwide vs. fiat that is specific to a sovereign nation. Crypto also has variety. Some crypto have volatile market values, while others have stable values pegged to stable assets.

Selected crypto in this book represents evolving technologies. Classifications are for the purpose of education and not investment advice. Main classification is by nth Generation (nG) for comparison of different types of ledgers for clearing and settlements, transaction verification, and cryptographic security methods.

Timeless

For example, 1G-3G crypto has blockchain ledgers with miners providing transaction verification, and progressive steps towards quantum security. Early crypto includes 1G Bitcoin (BTC), 2G Ethereum (ETH), and 3G Cardano (ADA). 4G-5G crypto has a variety of distributed ledgers with transaction verification that is green (i.e., Fast, Fee-less, and Miner-less (FFM)), along with further improvements in cryptographic security. Green crypto includes 4G IOTA (MIOTA), and 5G Nano (NANO) and Holo (HOT).

Now for more on what is beyond the blockchain.

F. BEYOND THE BLOCKCHAIN

A popular topic over the past decade has been the revolutionary blockchain technology. However, that technology has an energy consuming mining process for transaction verification. The next decade will likely include distributed ledgers with miner-less green methods for transaction verification. Green technologies would be more efficient for Internet Of Things (IOT), banks and social networks, sovereignty, transaction verification, and evolution to on-chain methods. *(Ref. F1)*

Ref. F1:
builtin.com/internet-things

1) Internet Of Things. The Internet and wireless communications created a new paradigm that allowed transactions between things with electronic identities—cars to maintenance facilities, refrigerators to shopping lists, and so on. The mechanics of IOT will come with green 4G-5G crypto.

4G IOTA has transaction verification via clients vs. miners, and single-entry Transaction ID (1xTXID) accounting. Verification will be with secure client software, similar to free https browsing. With 1xTXID, clients verify transactions with a settlement record on the Directed Acyclic Graph (DAG)-based Tangle distributed ledger. *(Ref's. F2, F3)*

5G crypto, such as Nano and Holo, have transaction verification via clients or agents vs. miners, with double-entry Transaction ID (2xTXID) accounting. Double-entry means there are separate transactions for clearing (i.e., a signed send/debit transaction) and settlement (i.e, a signed receive/credit transaction). Nano uses a DAG Block Lattice for parallel agent processing with each agent on a separate distributed ledger. Nano provides a scalable and secure foundation for financial transactions that can compete with Bitcoin. *(Ref. F4)*

Holo uses a Distributed Hash Table (DHT) for parallel processing with each agent effectively on a separate distributed ledger. "...each agent in the public blockchain maintains a private fork that is managed and stored in a limited way...with a distributed hash table..." Holo provides a scalable and secure foundation for applications and smart contracts that may compete with Ethereum. *(Ref. F5)*

Updates for modern crypto with green methods for transaction verification will be included in a *beyond-blockchain* website. *(Ref. F6)*

Ref. F2:
en.cryptonomist.ch/2019/06/30/dag-blockchain-use-iota/
Ref. F3:
iota.org/research/meet-the-tangle
Ref. F4:
mycryptopedia.com/nano-block-lattice-explained/
Ref. F5:
bitrates.com/news/p/holochain-part-2-how-ceptr-is-achieving-what-no-other-blockchain-was-able-to
Ref. F6:
jmwnuk.wixsite.com/digitalassets/beyond-blockchain

2) Banks and Social Networks. With billions of worldwide users, banks and social networks are creating new stablecoins with new methods for clearing and settlements. Their efforts can challenge aging ACH and RTGS in bank systems. Social networks can be competition for credit brands. Credit cards are in use 24/7 worldwide for shopping, restaurants, entertainment, and travel. Banks and social networks can modernize such that:

a) Fiat money is diversified with crypto such as Bitcoin and Ethereum, and stablecoins such as USDT, USDC, J.P. Morgan (JPM) Coin, and Facebook's Libra. *(Ref's. F7, F8, F9, F10)*

b) Fiat banks with RTGS's can diversify with modern settlements such as the Quorum DLT for the JPM Coin**.** *(Ref. F11)*

c) Fiat credit brands can diversify with Facebook's Novi (previously Calibra) financial services wallet on a smartphone. *(Ref. F12)*

Ref. F7:
coinmarketcap.com/currencies/tether/
Ref. F8:
coinmarketcap.com/currencies/usd-coin/

Ref. F9:
investopedia.com/jpmorgan-to-launch-jpm-coin-4587182
Ref. F10:
fortune.com/longform/facebook-libra-stablecoin-digital-currency-crypto/
Ref. F11:
goquorum.com/
Ref. F12:
theverge.com/2020/5/26/21270437/facebook-calibra-novi-rename-digital-wallet

3) Sovereignty. China has banned cryptocurrencies. At one time China had many traders and miners. What happened? With crypto, it was too easy to move money out of China. That lack of control threatened China's sovereignty. China understood the value of crypto, but needed better control of money flow. Their solution will likely be a sovereign stablecoin. Mu Changchun, a deputy director in the People's Bank of China (PBOC), is leading the effort for a sovereign crypto. That sovereign crypto could likely have value by decree, scalable capacity, with an added layer for a reputation rating. That rating could be more than a FICO® Score, and people and businesses could be rewarded (or punished) based on their rating. *(Ref's. F13, F14, F15)*

Ref. F13:
asiacryptotoday.com/
Ref. F14:
investopedia.com/terms/p/peoples-bank-china-pboc.asp
Ref. F15:
investopedia.com/terms/f/ficoscore.asp

4) Trace to Verified Source. Source is the beginning (i.e., genesis) of something of value. The genesis block in a blockchain or DLT is the first block. Chain means that every block, after the genesis block, references a previous block. That reference uses a cryptographic hash function to verify that a transaction has not been altered and can trace to a verified source. That trace to an unaltered and verified source is what allows transactions to be cleared and settled on a blockchain.

Timeless

For crypto, the blockchain/DLT process verifies transactions with financial value. That concept can also be used for non-financial applications--similar to United Parcel Service (UPS) tracking parcels. In a similar way, the "on-chain" process can track changes of title ownership, quality of parts used in construction transactions, food conditions from farm to grocery store, and medicine transactions from ingredients and formulas to pharmacy products.

Many organizations are providing education and building on-chain systems for tracking in non-financial applications. They include at least: Amazon, Deloitte, Harvard, International Business Machines (IBM), Microsoft, Massachusetts Institute of Technology (MIT), and Oracle.

5) Evolution to On-chain. Modern economies have off-chain and on-chain components. Off-chain includes systems with transactions that are not on a blockchain/DLT. Off-chain transactions can include those at home for lifestyle, healthcare and welfare; and what's trending for income, rewards and payments. Off-chain also includes existing financial systems for banking, investments, and insurance; along with a nation's transactions for its treasury and handling of assets and liabilities.

On-chain includes systems with transactions that are on a blockchain (1G-3G) or green DLT (4G, 5G). On-chain systems include crypto exchanges, crypto digital wallets, and crypto mining. Of particular interest are off-chain to on-chain trends for flexible **a)** income, **b)** rewards, and **c)** payments.

a) Income from payroll, social security, or welfare, can be deposited to crypto public address represented by a QR code on a modern debit card. That income can be sent by the corresponding on-chain organization. The recipient could choose if and when the crypto value is exchanged to a fiat value for traditional use with a modern debit card.

b) Rewards from shopping, dining, entertainment, or travel can be deposited with a process similar to **a)** above with value sent by the corresponding rewarding on-chain organization. The recipient's rewards card could operate as a modern debit card as in **a)** above.

c) Payment for travel and credit cards is defined in patent US10204378B1. With this patent, when it is time to pay credit card charges the consumer can make payments with money with best value-- either (off-chain) fiat or (on-chain) crypto. *(Ref. F16)*

With availability of off-chain and on-chain capabilities, users can choose which method has the best benefits. For example, fiat purchases with a credit card may have advantages at the point of sale with transactions confirmed in seconds. When it is time to pay those credit card charges, the best benefit would be a choice to pay with money with best value-- either fiat or crypto.

The words above can be used to create many pictures. One picture is Figure 1 as shown on the next page. The level of integration of off-chain and on-chain depends on the level of diversity in a nation's economy. More about Figure 1 is in **Appendix I. NEW CHOICES, Appendix II. NEW FOUNDATIONS, and Appendix III. NEW VOTING SYSTEM.**

Ref. F16:
patents.google.com/patent/US10204378B1/en

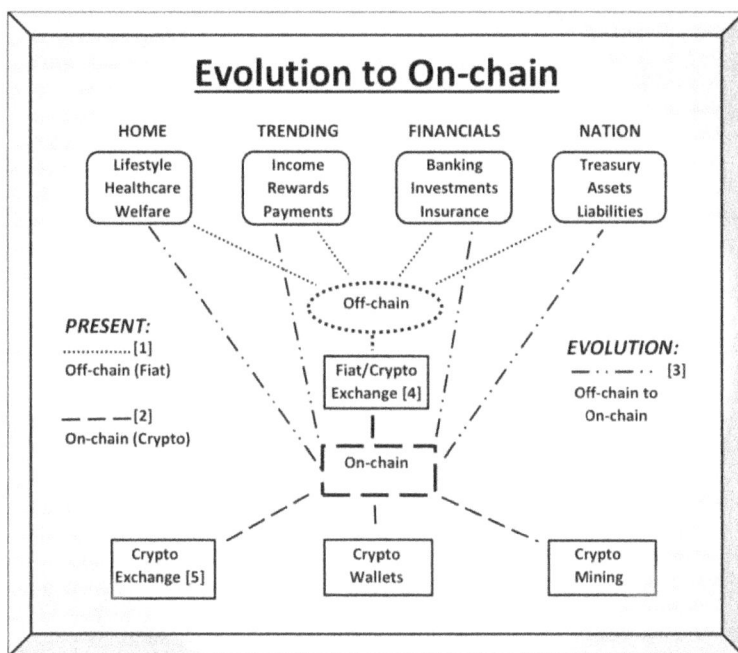

Figure 1.

[1] Off-chain: Transaction are not recorded on a blockchain or DLT.

[2] On-chain: Transactions are recorded on a blockchain or DLT.

[3] Evolution: Off-chain to On-chain: An off-chain economy can be modernized by diversifying with on-chain capabilities.

[4] Fiat/Crypto Exchange: An exchange, such as Coinbase, has fee-based capability to receive, store, and withdraw crypto, trade between different types of crypto, and trade between fiat and crypto.

[5] Crypto Exchange: An exchange that has fee-based capability to receive, store, and withdraw crypto, and to trade between different types of crypto, but does not have the capability to trade with fiat.

G. TOWARDS SUSTAINABLE ECONOMIES

1) Purpose. This section was written during the 2020 pandemic. At the time, shelter-in-place, social distancing and facial masks were the practice of the day--to avoid the Corona Virus Disease 2019 (COVID-19). That disease, also known as Wuhan Acute Respiratory Syndrome (WARS), was caused by the Severe Acute Respiratory Syndrome Corona Virus 2 (SARS-CoV-2). Because of the resulting worldwide deaths and economic damage, only time will tell if the spread of that virus was an accident or the start of a new form of biological and economic warfare. Lessons learned from the pandemic are that trade and money can be impacted by disease and viruses, and many changes are likely--due to the related economic impacts. *(Ref's. G1, G2, G3).*

> *Ref. G1:*
> *wsj.com/articles/china-rescinds-penalty-for-late-doctor*
> *Ref. G2:*
> *realmoney.thestreet.com/.../...we-ve-got-wars-to-worry-about-15214744*
> *Ref. G3:*
> *covid19.nj.gov/faqs/...can-covid-19-be-passed-via-currencies...*

What changes are needed? First could be to add digital diversity to modern financial systems. That change could complement fiat systems with digital methods to lower financial costs, reduce the spread of disease and viruses, and prepare for basic welfare for hardships or economic shutdowns. Digital methods could also have better IDs to verify transactions, reduce voter fraud (i.e., see **Appendix III**) and tax fraud, and reduce illegal and untraceable fiat transactions from counterfeiters and drug cartels.

2) Technology. Over the years much has changed with phones, TVs, and cars, but little has changed with fiat financial systems over the last half century. The fiat infrastructure is typically physical, expensive and tax-funded, with central banks for monetary policy, treasury for printing and minting money, expensive clearing and settlement systems for banking transactions, and aging postal mail. Also, fiat cash transactions are not traceable, and can be illegal or with counterfeit money. Is there an alternative that does not have these negatives?

Timeless

There is an alternative and it is not Apple Pay or other fiat-based credit/debit capability. A modern financial architecture could be based on genuine cryptocurrency (crypto) technology that includes digital wallets, blockchains or Distributed Ledger Technology (DLT), and 24/7 worldwide operation at Internet speeds. *(Ref. G4)*

A crypto digital wallet can operate as a bank or bank account. It is a paired public address and private key, with pairing through timeless methods of mathematics and cryptography. When the digital wallet is used as a bank, the owner of the digital wallet has the public address and private key. The owner of deposits to the public address, can withdraw value at any time using the private key. When a digital wallet is used as a bank account, a 3rd party such as Coinbase or Binance, owns the private key. That exchange "promises" to pay crypto value on request, just like a bank "promises" to pay fiat value on request.

For blockchains, much experience has been gained over the last decade with five generations of crypto technology. There is the 1st Generation (1G) Bitcoin and Blockchain, 2G Ethereum with a programmable blockchain, and 3G Cardano with parallel blockchains. Beyond the blockchain is green DLT with transaction verification that is fast, fee-less, and miner-less. Examples of green DLT are 4G IOTA's Tangle for the Internet of Things (IOT), and 5G Nano's Block Lattice and Holo's Distributed Hash Table (DHT). With green tech, effectively there is a distributed ledger for each user (i.e., client or agent) vs. one, easy to congest, blockchain.

Let us consider a future time. What if service levels of a legacy fiat economy could be retained or provided at better levels, but with a modern financial infrastructure. That infrastructure could have costs reduced by 10X or more, and address special needs—to reduce fraud and illegal transactions, and be prepared for economic shutdowns. The why for a change to a modern financial system may be obvious, but the how is not so obvious.

3) Systems, Currencies and Wallets. Modern digital systems, currencies, and wallets can address special needs in modern economies. A Smart Digital Wallet (SDW) could be a hybrid digital wallet to store income from a variety of sources, provide an exchange between crypto and fiat, and be used for credit and payments. The SDW could also have an ID with KYC/AML information for banking, investments, and voting, plus provide history for health and welfare needs. Examples that follow generally apply to the U.S., but could apply to any country that is evolving towards a modern economy.

The new job for the Fed or other private companies could be the production, storage, distribution, accounting, and security of possibly two types of new digital currency. One type could have stable value such as a Reserve Stable Cryptocurrency (RSC). A second type could have volatile market value such as a Reserve Market Cryptocurrency (RMC). There can also be competing private companies to manage policy for legacy fiat and modern crypto. Examples could be the Fed or Sovereign Reserve System-Fiat (SRS-F) for legacy fiat currency, and a Sovereign Reserve System-Crypto (SRS-C) for the stable RSC and market valued RMC.

An RSC could be equated by decree to a fiat $1 (or $10 or $100 or other). Eventually, just as the dollar became independent of gold, the RSC can become independent of the dollar--and related fiat debt. The RSC could be divisible to two places like a penny (e.g., if 1 RSC = $1, then 0.01 RSC = $0.01), or four places (e.g., if 1 RSC = $10, then 0.0001 RSC = $0.01), and so on.

The RMC would not have a fixed value like the RSC, but a value that can be volatile and depend on market supply and demand like Bitcoin (BTC).

For a time, fiat USD and crypto RSC and RMC systems could operate in parallel on a limited basis to debug process issues, evaluate cost and benefits, and fix errors or deficiencies. Expansion of crypto RSC and RMC systems with more features or to a larger population can occur as appropriate.

Timeless

Use of a SDW could be optional, and users could have control of what data is public or private. A parent or legal guardian could sign up a child at birth or anytime up to a predesignated age or default of 18. After that age, the individual could make their own choices unless legally restricted.

After sign up and registration, the individual could receive the SDW that could be in the form of an A/B pair of plastic cards the size of credit/debit cards. The A/B cards could be configured for a crypto/fiat currency pair for the country of use or residence. In the U.S. for example, the crypto/fiat pair could be BTC/USD or RSC/USD or as requested by the registered user.

The advantage of plastic cards is that they do not require power or communication like a smartphone or computer, but can be used anywhere that there is power and communications. Data with the A/B cards can be securely accessed through a registered user's computer or smartphone with user control of contents, security and backup.

The A card could have multi-use as a crypto SDW, and fiat credit or debit card, and include an ID for transaction verification. The A card could have a crypto public address and related optically readable Quick Response (QR) code, non-visible public and private keys for crypto send transactions, and computer chip(s) for operation with credit/debit reader or tap equipment. The SDW public address could be used for crypto deposits (i.e., income) from work, benefits, rewards, welfare, or from other SDWs. For world travelers, the fiat credit/debit card could be configured for world use, or the traveler may prefer two or more A/B pairs depending on preferred travel destinations—say a pair for the U.S. configured for RSC/USD, and a pair for the UK configured as BTC/GBP.

The A and B card owner could be able to control how deposits to the public address are used, from a registered computer or smartphone. As a default, income could be moved and converted to a fiat debit account. As a default override, crypto deposits can be stored in the SDW public address, sent to another SDW with a corresponding computer or smartphone app, or converted and used to pay fiat charges on a credit card as described in Flexible Payment Services. (*Ref. G5*)

The B card could have private information. It could include the public and private keys used for crypto withdrawals from the corresponding A card. The B card is what makes the SDW operate like a bank. The user could have control of all data in the SDW, along with KYC/AML information that could be part of an ID for access to fiat banking and investments accounts.

Health, welfare or other information that the individual prefers to be private could also be on or linked to the B card. Imagine that instead of a 10 page information form for the next doctor visit, all the health information that is needed could be accessed through the B card.

In case of a pandemic or economic shutdown, past records could be available to determine basic welfare incomes. Basic incomes could be calculated based on previous experience and conditions. If an individual should get RSC valued at $X per week for Y weeks, there would be no need for delays in congressional debates.

Basic incomes for companies could also be calculated for different conditions and circumstances, along with avoiding the bail-out of mismanaged or shell organizations that need to be restructured or closed. *(Ref. G6)*

Ref. G4:
pocket-lint.com/...what-is-apple-pay...and-which-banks-support-it
Ref. G5:
patents.google.com/patent/US10204378B1/en
Ref. G6:
cbsnews.com/news/how-shell-companies-launder-dirty-money/

4) Challenges. Key challenges will be public acceptance and education. Public acceptance may be accelerated if there are significant benefits in the use of SDW vs. fiat. Retraining bureaucracies associated with sovereign fiat currency could also be a challenge. Again, significant benefits with handling SDW vs. fiat can be the key to acceptance.

Other challenges in use of SDW could be learning the crypto process and recourse for problems. Section D starting on page 10 of this book has more on the crypto process, and recourse for lost, destroyed or stolen SDW could be similar to Section D, part 9), pages 17-18. Additional instructions can be provided for best practice in setting up secure accounts, and preventing errors in a purchase or payment process.

Lost or destroyed SDW's could be covered as part of the initial sign up and registration, that could include instructions for backup and replacement. Temporary backup with a smartphone app could be available until replacement cards are received.

Stolen SDW's could be blacklisted to prevent further use. For illegal use before being blacklisted, transactions could be traceable and stores may be able to record pictures of users of lost or stolen SDW's. Methods could be in place to discourage illegal use of SDW's, such that costs would exceed potential gains from illegal or harmful actions.

5) Jobs. Many any new jobs will be created with use of SDW. Examples include SRS-F and SRS-C administration, support, and R&D; A and B card design, development, test, manufacture, default configurations, distribution, security, and support; and RSC and RMC production, storage, distribution, accounting, and security.

Payment for these new jobs could come from fiat or crypto (i.e., RSC and RMC). Workers could have a choice of being paid in fiat or crypto. For crypto, there could also be a choice between stable RSC or market valued RMC. Payments could also be split over fiat and crypto.

6) Digital Transformation. Intent of digital transformation is that evolution of trade and money helps to create a safe and healthy environment for people to live, work, and play. Evidence of progress towards digital transformation could be observed in **a)** verifiable transactions, and **b)** sustainable economies.

Timeless

a) Progress towards verifiable transactions includes:
1. Methods to verify that data sources are valid and message content has not been altered during transfer from source to destination;
2. Distributed ledgers of verified transactions are used to eliminate voter fraud and tax fraud, and cause demise of counterfeiters and drug cartels;
3. Supply chains with traceability to cause of problems; and
4. Alerts for false news or advertisements.

b) Progress towards sustainable economies includes:
1. Verifiable transactions as in **a)** above;
2. Effective and efficient reserve systems and currencies;
3. Lower costs and quantum security in voting transactions;
4. Smart digital wallets for income, credit, payments, and identity;
5. Tolerance to pandemics and related economic shutdowns;
6. Basic incomes for welfare and hardships; and
7. Financial systems that do not spread disease and viruses.

Appendix I. NEW CHOICES

"The future is not what it used to be." An interesting quote attributed to several persons including Yogi Berra, Paul Valery, and others. As shown in Figure 1, off-chain was the world economies before crypto. Much can be learned from off-chain progress that can be applied to on-chain innovations. *(Ref. I1)*

Ref. I1:
quoteinvestigator.com/2012/12/06/future-not-used/

1) Off-Chain Progress. Learning from off-chain progress, of particular interest, would be in **a)** communication systems and their scalability for growth, and **b)** fiat money processes that are tangible and intangible.

a) Observables in a communications system can be compared with an iceberg. The tip of the iceberg is visible but most is unseen and below the water. In communications, what starts as the tip with voice, data, or video on visible devices such as a phone, computer or TV, quickly becomes unseen bits passing over unseen physical facilities such as modems, Wi-Fi routers, cables, Telco or Internet Service Provider (ISP) facilities, and then on to a voice or data destinations on the World Wide Web (WWW). *(Ref's. I2, I3)*

Ref. I2:
whatismyipaddress.com/isp
Ref. I3:
techterms.com/definition/www

Progress in mobile wireless communications has evolved over several decades from 1G Analog with poor quality and dropped calls, to digital with high quality and performance. Digital wireless communications technology included 2G Time Division Multiple Access (TDMA) with the signal on one frequency that is time slotted to serve multiple users, to 3G Code Division Multiple Access (CDMA) with more security, to modern Orthogonal Frequency Division Multiplexing (OFDM) with the signal carried on multiple frequencies that are time slotted for more users and higher speeds. OFDM is the current technology for Wi-Fi, 4G mobile wireless, new deployments of 5G mobile wireless, and planned future deployments of 6G mobile wireless communications. *(Ref's. I4, I5, I6)*

Ref. 14:
yourdictionary.com/tdma
Ref. 15:
yourdictionary.com/cdma#computer
Ref. 16:
yourdictionary.com/ofdm#computer

Scalability in wired and wireless communications services is the ability to adapt to changes in the location and number of users and their voice/data/video traffic. Scalability in communications systems has come from the technical innovation of a Recursive Inter-Network Architecture (RINA). A basic premise of RINA is that networking is inter-process communications (IPC) and different systems can communicate and share data using a Distributed IPC Facility (DIF). Without scalability through RINA, instant and quality communications would not exist. With RINA, instant dial tone on wired voice calls, and rare outages for wired or wireless message, data and video services are common. *(Ref's. 17, 18, 19)*

Ref. 17:
riverpublishers.com/pdf/ebook/chapter/RP_9788793519114C16.pdf
Ref. 18:
techopedia.com/definition/3818/inter-process-communication-ipc
Ref. 19:
queue.acm.org/fullcomments.cfm?id=2076798

b) Many think of money as the tangible part that is in their wallet or purse. But in reality most money, like the hidden parts of a communications system, is an intangible number in a credit, bank or investment account.

While many like the idea of having pocket money for shopping, most fiat money transactions are not with tangible cash but with an on-line Internet purchase, or at a store or gas station with a credit/debit card. When more understand that most money is an intangible number and not tangible paper and coins, then they become ready to learn more about safe and secure on-chain innovations--with lower costs, faster transactions, and built-in protections against fraud and counterfeiting.

2) On-Chain Innovations. On-chain first appeared in 2009 with a Bitcoin money layer on the Internet. Secure transactions between peers bypassed banks and costly delays of wire transfers. After a decade of growth, industry and governments can now see the benefits of on-chain methods for financial and non-financial applications.

Benefits of on-chain methods for financial applications includes performance (e.g., clearing in minutes vs. days for ACH), reduced costs (e.g., distributed ledgers vs ACH, RTGS and wire transfers), prevention of double spends (i.e., counterfeits), and tracing to source of illegal transactions vs. untraceable fiat transactions. While evolution to modern economies as described in Section G could take decades, fiat and crypto synergy opportunities can soon be underway. Consumers can keep the fiat tangibles they like such as paper money, coins and checks, but costs of the fiat infrastructure can be significantly reduced with distributed ledgers initially replacing ACH, RTGS and wire transfers.

For non-financial applications, on-chain methods can provide improvements to supply chains. With current off-chain methods, efficient and timely tracking to sources of defects in supply chains is unlikely. Methods for tracking and tracing to defects can be unified with on-chain methods for: **a)** food, **b)** gems and minerals, **c)** medicine, **d)** equipment, **e)** news, **f)** photoshopping, and **g)** IDs and voting.

a) If a food product is bad and removed from store shelves, and the process from production to delivery is tracked, quick identification of the source of defects can result in timely repair. Considering coffee, for example, it would be important to know if expensive coffee came from the claimed legal source that was environmentally friendly and not full-sun coffee beans.

b) If the source of gems was traced to blood diamonds, that tracing could result in reduced sales and/or improved conditions for production without child labor. The same could apply for the new blood mineral cobalt. Cobalt is used in jet engine alloys, as a medical tracer in radiotherapy for cancer, and now has greatest use in lithium batteries for smartphones, computers and electric vehicles.

c) If a medicine is defective and causes injury or death, tracing back to sources can enable the understanding of the cause of the defect to be followed by ingredient or formula improvements.

d) If brakes or other equipment on a vehicle are defective, fast tracing to sources and quick repair or replacement can save lives.

e) News is data that is an essential part of an economy. News that is not based on facts, but on biased opinions or false witness, can lead to bad choices for a misinformed public and businesses. Bad choices can eventually lead to disaster for an economy. Fact-based news that can be traced to verified sources will likely be part of modern healthy economies.

f) Photoshopping is altering a digital image. Popular uses include removing age marks to improve appearance, retouching old photos to make them look new, or adding color to a black and white image. Instructional videos show how to make the changes. The darkside is that digital editing has been used for false advertising. An example could be a famous person falsely endorsing a diet pill or brain-boosting pill. With modern tracking methods, consumers can then be warned when an image has been altered without permission.

g) IDs are essential for voting, licenses, and more. Imagine the disaster if anyone could vote with a fake ID, or if votes can be manipulated by biased third party handlers. IDs traced to verified birth date and time, location, parents, and more can be the norm. Votes not counted or more votes than legal voters are signs of fraud. The voting process must be secured so that only valid voters can vote, and voters can then verify their votes were counted towards designated recipients. *(Ref. I10)*

Ref. I10:
https://...cointelegraph.com...south-korean-gov...to-bring-blockchain-voting...amp

Making a choice to diversify with what is new can be a slow process. It took decades for the value of the (off-chain) Internet to be understood. Use of the Internet has become common for mail, messages, wireless phone calls, streaming TV, web browsing, social networking, and on-line purchases and payments. In a similar way, understanding new on-chain innovations and their foundations can give insight to potential benefits in financial and non-financial applications.

Appendix II. NEW FOUNDATIONS

What follows is more on new foundations for the on-chain world to include: forks in the road, scalable capacity, quantum security, and importance of learning from dark and bright sides of history.

1) Forks in the Road. Forks are part of crypto culture. They are similar to software upgrades in the fiat world—such as regular bank maintenance. Crypto, such as Bitcoin and Ethereum have roadmaps for modernization that includes schedules for upgrades. Points in time for upgrades are called "forks." Planned forks can be for new features, improved performance, or reduced costs. Unplanned forks can also occur to repair defects. Forks are also classified as **a)** Soft or **b)** Hard. *(Ref's. II1, II2)*

Ref. II1:
medium.com/swlh/hard-forks-and-soft-forks...5
Ref. II2:
visualcapitalist.com/major-bitcoin-forks-subway-map/

a) Soft forks are new protocol rules for a blockchain with backward compatibility. Soft forks do not result in splitting the main chain or a new cryptocurrency--as can occur with a hard fork. Backward compatibility with soft forks means that all mining nodes do not need to immediately upgrade to new rules. Older nodes that have not upgraded can continue viewing transactions and mining new blocks. However, older nodes risk trying to push invalid blocks onto the blockchain, that will be rejected by the upgraded nodes. This results in loss of computational resources, and forces all nodes to eventually upgrade to the new protocol rules.

A soft fork example is implementation of Bitcoin's SegWit (Segregated Witness). This fork more than doubled transaction capacity in a block by removing the public key and signature and adding them into a separate channel. The goal of the soft fork was to solve Bitcoin's low transaction rate, due to a small block size and 10 minute block times. SegWit allowed older nodes to continue creating the old type of blocks, but also allowed upgraded nodes to create the new and more-efficient blocks. *(Ref. II3)*

Ref. II3:
investopedia.com/terms/s/segwit-segregated-witness.asp

b) Hard forks are new protocol rules for a blockchain that is not backwards compatible with the old rules. All nodes must reach full consensus to the new rules if the blockchain is to continue as a single chain. Otherwise, the blockchain will be split into a second chain. An example of a hard fork that received full consensus was Ethereum's Istanbul update that included several Ethereum Improvement Proposals (EIPs) at the same time and resulted in a significantly improved Ethereum network. *(Ref. 114)*

When a hard fork does not receive full consensus, there is a splitting of the blockchain into separate chains along with creation of a new coin. A classic hard fork that split the blockchain was Bitcoin (BTC) to Bitcoin Cash (BCH) with a larger block size for more capacity. When BCH was created, it came with a similar number of coins as BTC. So if a Bitcoin wallet had 10 BTC at the time of the fork, the owner of that wallet would also get a separate Bitcoin Cash wallet with 10 BCH. The BCH coin would have separate market value than BTC depending on demand. This type of hard fork is very different from the fiat world that has no equivalent in the sense of doubling the number of owned coins. *(Ref. 115)*

Another hard fork example was Ethereum (ETH) splitting to Ethereum Classic (ETC). The battle was due to the immutability concept. Many ETH leaders and miners wanted to correct and invalidate an ETH Digital Autonomous Organization (DAO) hack worth millions vs. a contingent of miners that did not want to invalidate the hack. After review and debate, the majority of ETH persons decided to go back in the chain to a block before the DAO hack with a hard fork at ETH block 1,920,000.

The ETC hard fork continued with the DAO hack of ETH coins. However, the new ETC chain was not affected by the ETH hack, since ETC coins were not hacked. This choice of separate actions by the ETH and ETC communities was significant to the crypto world. It demonstrated that "justice" could be achieved for the ETH community that wanted to void the DAO hack, and "truth" could be obtained for the ETC community by retaining the "immutable" characteristic of a crypto. *(Ref. 116)*

From a user perspective, if an Ethereum wallet had 5 ETH at the time of the hard ETC fork, the owner of that wallet would also get 5 ETC in a separate Ethereum Classic wallet. ETC would have separate market value from ETH depending on demand.

So far, new coins created from a hard fork typically have less market value than the original coin. While BTC has typically been valued in thousands of $'s, BCH is typically hundreds of $'s. Similarly, ETH is typically in the hundreds of $'s while ECH is typically in the tens or more likely units of $'s.

Ref. 114:
cointelegraph.com/magazine/ethereum-hard-fork-istanbul-2019/
Ref. 115:
investopedia.com/tech/bitcoin-vs-bitcoin-cash-whats-difference/
Ref. 116:
currency.com/ethereum-vs-ethereum-classic-the-difference

2) Scalable Capacity. Scalability is the ability to change capacity for changes in users and their locations and traffic. Scalability is essential for sustainable clearing and settlements in financial transactions, and tracking and tracing in non-financial transactions. Scalability must also address trade-offs with security and decentralization. Early Bitcoin and Ethereum protocols did not have scalable capacity for clearing and settlements. They may be great for store of value, but too slow at the point of sale compared to credit/debit cards. Miners have a severe bottleneck when they try to get their transactions on a blockchain. Less than 10 transactions a second (tps) for Bitcoin and slightly more for Ethereum will never be able to compete with more than 1,000 tps with credit and debit cards.

The year 2020 will likely be the start for significant improvements in scalable capacity with **a)** lightning networks in Bitcoin, **b)** sharding in Ethereum, **c)** scalability in Cardano, and **d)** scalability in IOTA. *(Ref. 117)*

Ref. 117:
hackernoon.com/the-blockchain-scalability-problem...4

Timeless

a) The Lightning Network is intended to provide scalable capacity for Bitcoin. Development is part of BIP0068 and BIP0112. An off-chain (or side-chain) payment channel between peer users will start and end with on-chain transactions that are part of a smart Hashed Timelock Contract (HTLC). FFM off-chain transactions will be allowed while the payment channel is open.

For security, the smart HTLC will require Bitcoin deposits for the payment channel setup. If one of the users becomes a bad actor (i.e., scammer, hacker) and does not honor the smart contract, their deposit will be withdrawn and given to the other peer user.

The idea is to stop hacking attempts when the cost of a hack would far exceeds any possible return. Some of the features in the Lightning Network proposal include: rapid payments, no third-party, reduced blockchain load, and channels that stay open indefinitely. More details are at *Lightning_Network. (Ref. II8)*

Ref. II8:
en.bitcoin.it/wiki/Lightning_Network

b) Sharding is the Ethereum approach to scalability. A blockchain database would be split into several parts called "shards," with parallel processing in each shard as the key to scalability. Sharding is expected to be implemented as part of Ethereum 2.0 starting in 2020 with six phases corresponding to soft forks including: basic sharding, state transition, light client, cross-shard transactions, tight coupling, and exponential sharding. Details for each phase can be found at *Sharding-roadmap. (Ref. II9)*

Ref. II9:
github.com/ethereum/wiki/wiki/Sharding-roadmap

c) Scalability in Cardano will follow techniques similar to what was perfected in mobile wireless communications including RINA, TDMA and OFDM. To handle a large and growing number of transactions, Cardano splits the network into smaller "subnetworks" using a RINA technique. Each node can become part of a subnetwork and communicate with other nodes in other subnets.

Timeless

Examples of use the of RINA in mobile wireless communications could be a large stadium with thousands watching a sporting or entertainment event. Spectators with smartphones could be texting, talking and streaming video to others worldwide. How all that works with fast setup and take down of capacity is RINA technology, and Cardano is bringing RINA-like implementation to the world of cryptocurrencies.

Epochs in Cardano are similar to TDMA and OFDM perfected in mobile wireless communications. Time in Cardano is divided into epochs. Every epoch is divided into slots. Every 20 seconds there is a slot and an epoch contains 21,600 slots. For scalability, the number of slots per epoch can be changed and there can be parallel epochs. In comparison with mobile wireless, operation with one epoch is like TDMA and operation with parallel epochs is like OFDM. More on Cardano scalability can be found at *basho. (Ref. Il10)*

Ref. Il10
cardanoroadmap.com/en/basho/

d) Scalability in IOTA will be achieved with client verification of transactions vs. miner verification. As the number of client users increases, so does the verification capacity. The technique for transaction verification is based on DAG and Coordicide. In DAG, the letter "D" (i.e., Directed) implies a start and end point, such as a destination in navigation or source of value in a financial transaction, "A" (i.e., Acyclic) means the start and end routing cannot get stuck in an endless cyclic pattern, and "G" (i.e., Graph) is a popular part of mathematics that is finding a modern use in crypto applications.

Coordicide is the decentralized version of IOTA's older centralized coordinator. It provides scalability in terms of transactions per second, and has the ability to create smart contracts. The whitepaper is at *Coordicide_WP.pdf,* and there is an interesting visual model of the IOTA DLT at *tangle.glumb. (Ref's Il11, Il12)*

Ref. Il11:
files.iota.org/papers/Coordicide_WP.pdf
Ref. Il12:
tangle.glumb.de/ (Caution. This is a http website and not secure https.)

3) Quantum Security. On-chain foundations must be tolerant to hackers with quantum computers attempting to decode private keys. Quantum computers are super computers rated in quantum bits (qubits). In classical computers, each bit can be in one of two states—on or off, or 0 or 1. In quantum computers, probabilities come in to play, such that a theoretical qubit can have more than two states. In general, a qubit rating of 1 implies capability of present computers, and qubit = n implies 2^n capability over present computers—less error correction to deal with the statistical nature of quantum computers. *(Ref. II13)*

Current claims are qubit ratings near 50 from IBM and Google, and 2,000 from D-Wave Systems, Inc. One hopes that this powerful qubit capability will be used to help determine causes of cancer, alzheimers and other serious diseases; and not decoding of private keys or logins and passwords to steal value and personal data. *(Ref's. II14, II15)*

Ref. II13:
scientificamerican.com/observations/the-problem-with-quantum-computers/
Ref. II14
cnet.com/news/ibm-now-has-18-quantum-computers-in-its-fleet-of-weird-machines/
Ref. II15:
techcrunch.com/2019/11/15/d-wave-sticks-with-its-approach-to-quantum-computing/

Crypto professionals have a variety of security methods that are tolerant to hack attempts from quantum computers. In the years 2020 and later there will likely be enhanced quantum security in many areas including voting (i.e., see **Appendix III**). Methods for quantum security include: elliptic curves, sponge construction, Hierarchical Deterministic (HD) wallets, and more advances in Public Key Cryptography (PKC) as described in **a)** through **e)**. *(Ref's. II16, II17)*

Ref. II16:
investopedia.com/terms/h/hd-wallet-hierarchical-deterministic-wallet.asp
Ref. II17:
techopedia.com/definition/9021/public-key-cryptography-pkc

Timeless

a) PKC with ECDSA, secp256k1's elliptic curve y^2 =x^3 + 7, and SHA-256 hash is used in Bitcoin Wallets. Quantum hackers focus on ECDSA as it is a National Security Agency (NSA) standard, with decode of the private key estimated by 2030 or earlier. *(Ref's. II18, II19, II20, II21)*

Ref. II18:
en.bitcoin.it/wiki/Elliptic_Curve_Digital_Signature_Algorithm
Ref. II19:
en.bitcoin.it/wiki/Secp256k1
Ref. II20:
en.bitcoin.it/wiki/SHA-256
Ref. II21:
nsa.gov/

b) PKC with ECDSA, secp256k1 and the Keccak-256 hash based on sponge construction is not an NSA standard and is used in Ethereum Wallets. Sponge-based construction can compress or expand a bit stream by taking an input bit stream of any length and produce an output bit stream with a desired length. *(Ref's. II22, II23)*

Ref. II22:
quora.com/...difference-between-SHA-256-and-Keccak-256...
Ref. II23:
keccak.team/sponge_duplex.html

c) PKC with EdDSA, elliptic Curve25519 (Ed25519) and SHA-512 hash is not an NSA standard and is used by Nano and Holo, and in Cardano's HD Daedalus Wallet. Daedakus is Cardano's wallet for the ADA cryptocurrency. Use of Ed25519 and HD Wallets, hides the wallet that has value and could add years or decades to private key decode time by quantum hackers. *(Ref's. II24, II25, II26, II27)*

Ref. II24:
tools.ietf.org/html/rfc8032
Ref. II25:
ed25519.cr.yp.to/
Ref. II26:
computerlanguage.com/results.php?definition=sha512
Ref. II27:
cardano.org/en/the-daedalus-wallet/

d) PKC with a Winternitz DSA and Troika hash is used in IOTA's Trinity Wallet. The Troika hash, based on sponge construction, was designed by CYBERCRYPT to tolerate all known attacks from crypto and quantum hackers. The (Robert) Winternitz DSA builds on the work of Leslie Lamport and Ralph Merkle, and groups w bits to be signed with n groups for each wallet. With this approach, the user can vary w and n in time to make speed or storage the priority depending on load conditions and security objectives. *(Ref's. 1128, 1129, 1130)*

Ref. 1128:
patents.google.com/patent/US8386790B2/en
Ref. 1129:
cyber-crypt.com/troika/
Ref. 1130:
trinity.iota.org/

e) Crypto exchanges also hide the wallet with value by using Exchange wallets. Value is moved from an owner's on-chain electronic wallet to an Exchange wallet. An exchange, such as Coinbase, also offers off-chain "vault" services for added security. A withdrawal from a vault requires 2FA, two co-signers to approve the withdrawal, and a 48 hour delay before the withdrawal occurs. *(Ref's. 1131, 1132)*

Ref. 1131:
bitcoin.stackexchange.com/questions/56696/off-chain-transactions-with-coinbase
Ref. 1132:
coinbase.com/vault

One wonders if it could ever be possible for the quantum hacker to decode a private key or hack a financial account? Their time and effort would have more value if used to help determine causes and cure diseases, poverty, and illiteracy.

4) Abraxas. No discussion about the evolution of trade and money can be complete without a high level review of the **a)** dark side including Mt. Gox, Silk Road, and a 51% attack, and **b)** bright side including growth, verifiability and sustainability. It is important to learn from both--to leverage the positives and not repeat the negatives. *(Ref. II33)*

Ref. II33:
genies.fandom.com/wiki/Abraxas

a) Mt. Gox was a crypto exchange that went bankrupt as a result of thousands of Bitcoins being hacked. It is an example of what can happen when a crypto exchange is poorly managed and does not follow sound financial practice. *(Ref. II34)*

Silk Road was a crypto operation that allowed transactions of illegal products. The darker side of Silk Road was that investigating government agents were stealing confiscated Bitcoins. *(Ref's. II35, II36)*

A 51% attack could destroy a crypto system (or nation as described on page 17). A 51% crypto attack occurs when a mining pool controls more than 50% of mining resources. Dark side impacts could include transactions not getting confirmed, and confirmed transactions getting reversed. Self-monitoring has been the prevention approach. By observing a possible 51% attack, the crypto community can alert or black list the offending operation. *(Ref. II37)*

Ref. II34: Mt. Gox:
techcrunch.com/2019/02/06/the-plot-to-revive-mt-gox.../
Ref. II35: Court Case against Ulbricht (aka Dread Pirate Roberts):
caselaw.findlaw.com/us-2nd-circuit/1862572.html
Ref. II36: Court Case against Agents Force and Bridges:
justice.gov/.../2015/03/30/criminal_complaint_force.pdf%20%20%20
Ref.II37: Monitored Mining Pools to prevent a 51% attack:
blockchain.com/pools?timespan=4days

b) On the bright side, there are many examples. First is the growth of crypto from Bitcoin in 2009 to thousands of Altcoins in 2020 with billions in value. Clearing and settlements with crypto are more economic, faster and more secure than fiat. The founders of cryptocurrencies are modern heroes and are the pioneers of innovations of tomorrow. *(Ref's. II38, II39)*

Timeless

Verifiability can make money laundering a thing of the past with traceability to verified sources. Traceability will also have application in non-financial applications. Trace to source of defects can have priceless health and safety benefits. Sustainability becomes evident with crypto and DLT that are effective and efficient. The modern era will include 4G-5G crypto and communication foundations for IOT, AI, and quantum computing. *(Ref. II40)*

Read *Blocky explains Blockchain* and then read it to your grandchildren. If you have a unique story to tell, do it and self-publish. Instructions are provided to help get that important book started. *(Ref's. II41, II42)*

Ref. II38: Blockchain use cases:
builtin.com/blockchain/blockchain-applications
Ref. II39: Founders and creative pioneeers of tomorrow:
<> Bitcoin (BTC)--Satoshi Nakamoto: *en.bitcoin.it/wiki/Satoshi_Nakamoto*
<> Cardano (ADA)--Charles Hoskinson: *iohk.io/en/team/charles-hoskinson*
<> Ethereum (ETH)--Vitalik Buterin: *cointelegraph.com/...vitalik-buterin*
<> Holo (HOT)--Arthur Brock: *thedailychain.com/...arthur-brock/*
<> IOTA (MIOTA)--David Sonstebo: *cryptoslate.com/.../david-sonstebo/*
<> Nano (NANO)--Colin LeMahieu: *list.wiki/Colin_LeMahieu*
<> Ripple (XRP)--Chris Larsen: *forbes.com/.../chris-larsen/#66494a2057ed*
Ref. II40: Beyond Blockchain for the IOT, AI, and quantum era:
jmwnuk.wixsite.com/digitalassets/beyond-blockchain
Ref. II41: Blocky explains Blockchain--A kids' book about the blockchain:
amazon.com/dp/1775324222
Ref. II42: Instructions to self-publish a book on a favorite topic:
<> *kdp.amazon.com/en_US/*
<> *archwaypublishing.com/*
<> *bowker.com/products/Self-Publishing-Solutions.html*

Appendix III.   NEW VOTING SYSTEM

Traditional voting has manual handling, postal mail and paper ballots, and scanning and recording machines with questionable quality and performance. Manual handling and questionable machines can be likely sources of unintentional errors or intentional fraud. The US Postal Service (USPS) has patented a way to improve traditional voting with integration of blockchain technology, but continues to rely on manual handling, postal mail, and paper ballots. (*Ref. III1*)

Claims that traditional systems are secure with anonymous voters is flawed. There is no way to check a mailed ballot is legal and authentic after the envelope, supposedly with a validated signature, is removed from the ballot. Once that envelope is removed, it could be impossible to later determine if a ballot was legal and came in an envelope, or was fraudulent and added by an unknown source. Voting in person with a Direct Recording Electronic (DRE) machine can also be flawed. There is no way to quickly check if the recorded ballot was changed by the machine. Voting will likely continue with traditional methods until nations understand the negative consequences and start to phase in modern methods—that are more secure, accurate, reliable, and faster and less costly than traditional methods.

Ref. III1: *USPS Secure Voting System:*
https://patents.google.com/patent/US20200258338A1/en

1) Q-Blockmatrix. The new voting system will be referred to as "Q-Blockmatrix." Important features of this new system are that: **a)** voters will be pseudo-anonymous, **b)** only the voter, and not a third-party person or device, can validate that their ballot was recorded as intended—without errors or unauthorized changes, and **c)** there is a dashboard that can trigger action when there is a suspected error or fraud.

If there is a suspected error or fraud, that's when it gets challenging and what follows addresses that challenge.

Components of Q-Blockmatrix (QB) include:

- Q-Blockmatrix.com (QB.com) public ledger of voter authorization tokens, with corresponding
- Q-Blockmatrix.info (QB.info) private clearing database (i.e., for unmarked and marked but unconfirmed ballots), and public settled database (i.e., for marked and confirmed ballots) and their summaries.
- Q-Blockmatrix.xyz (QB.xyz) private database of confirmed fraudulent ballots, and a
- Q-Blockmatrix.site (QB.site) private dashboard to help detect suspected errors or fraud, and a private database for voting instructions and trouble tickets.

The QB.com ledger provides a record of transactions of voter authorization tokens, with corresponding ballot detail in QB.info. If any ballot data in QB.info is suspected to be an error or fraud, it would be flagged as such and be private. If a suspected error in QB.info is confirmed to be unintentional, the error would be fixed and the flag removed. If suspected data is confirmed to be intentional fraud, the flag would remain and the fraudulent data would be copied to QB.xyz and used as evidence for subsequent legal warrants and prosecution.

The new voting system would have flexibility for use in a variety of voting events and countries. It would be voter friendly and based on modern electronic technologies. Each voting event can be for a particular country, time, geographic region, candidates, and propositions. For the USA, geographic regions could be voting districts (i.e., precincts and wards), townships, cities, counties, states, and the nation. There would be a central voting authority for each voting event, and each voting district would have a voting authority that manages the voting process for that district's jurisdiction. Voters get registered in a particular voting district and vote for candidates and propositions on their ballot.

Candidates and propositions would depend on the time and region of the voting event. QB can operate in parallel with traditional methods during test and early use—as a way to compare quality, performance and cost. For other countries, QB could be tailored to their unique voting needs.

Timeless

2) Voter Perspective. Voters would use a computer or smartphone to vote from home, work, or any travel location that has access to electronic communications. Voters that do not have a computer or smartphone, would receive postal mail with instructions for special voting. Considering costs and effectiveness of a modern electronic system plus special capabilities for those that are electronically challenged, compare with traditional costs and effectiveness for a national election with hundreds of facilities for DRE voting, hundreds more gymnasium-sized facilities filled with thousands of people manually handling basket loads of envelopes and paper ballots, along with hundreds of scanning and recording machines with questionable quality and performance.

For a legal registered voter using QB, the process starts with instructions. The voting authority would email to each registered voter a secure access link to documentation on how to vote with a computer or smartphone. The instructions would include how to: **a)** get started with a digital wallet, voter authorization tokens, and secure access to a voting ballot, **b)** enter a voter authorization token to the ballot before marking the ballot with choices, and when completed view a unique receipt for the completed ballot, **c)** email the voting authority when voting has completed (i.e., cleared), and **d)** after the voting authority returns email that their validation has completed, validate again that the ballot was recorded as intended--without errors or unauthorized changes (i.e., settled).

Note in **d)** that only the registered voter validates their ballot--and not unknown third-party persons or devices. After clearing and settling the ballot, the voting process would be complete--unless there is a suspected error or fraud. If there is suspected error or fraud, the process in **5)** would be followed.

3) Electronic Technologies. The new voting system would use existing electronic technologies. It would integrate a distributed ledger technology such as blockchain with modern computers, smartphones, software applications, and electronic communications. Voting is paperless--other than for users printing copies of screen shots or other material, and voters without a computer or smartphone that would receive postal mail with special instructions.

Timeless

There would be three types of electronic communications with the new voting system: **a)** computer or smartphone "email," **b)** website "hypertext," and **c)** cryptographic "send." **a)** Email would be for distribution of secure links to instructions or trouble tickets. (Message service would not be used due to limited security.) **b)** Hypertext allows website access with a simple click on a secure hotlink. Security for website access would be with existing protocols including Transport Layer Security (TLS) and Hypertext Transfer Protocol Secure (HTTPS). (*Ref's: III2, III3*) **c)** Send would be for secure communications using Public Key Cryptography (PKC) with transaction records on a distributed ledger.

PKC comes with digital wallets—as used in cryptocurrencies with permanent transaction records on a distributed ledger or blockchain. A digital wallet is a public address and private key. Digital wallets allow clearing of transactions in minutes vs. days for clearing financial transactions in banking systems, and settlement of confirmed transactions soon thereafter. A voter authorization token would be a number in a digital wallet—a "1" means one voter authorization token, a "2" means two authorization tokens, and so on. The combination of TLS, HTTPS and PKC allows secure communications over possible non-secure facilities, such as mobile wireless or the Internet. (*Ref's: III4, III5, III6*)

As described on pages 44-46, PKC has several methods for security, intended to prevent hacking from quantum computers. The PKC method for the new voting system could be kept proprietary to help prevent fraud. PKC has five generic steps to send a message, such as a voter authorization token, from a sender's wallet to a (peer) receiver's wallet:

1) Key and Address Generation (random method -> private key -> public key -> public address),
2) Signing Algorithm (private key, message, digitally signed message),
3) Send Transaction (public key, digitally signed message, message),
4) Verification Algorithm (public key, digitally signed message, message),
5) Pass/Fail of Send Transaction.

"Pass" of a Send Transaction means the message source is valid and was not altered in the transfer, and the message can be used in subsequent processing. "Fail" means the message source may not be valid, or the message was altered in transfer and is not useable.

Note from 3) Send Transaction and 4) Verification Algorithm above, that PKC does not reveal the private key of the sender. (*Ref's. III7*)

QB has ten levels (i.e., *I* through *X*) of security as discussed in **4)** below, and together with TLS, HTTPS, and PKC will be referred to as "quantum secure voting."

Ref. III2: Transport Layer Security (TLS):
https://www.techopedia.com/definition/4143/transport-layer-security-tls
Ref. III3: Hyper Text Transfer Protocol Secure (HTTPS):
https://www.techopedia.com/.../hypertext-transport-protocol-secure...
Ref. III4: Mobile Wireless Communications:
https://www.slideshare.net/mobile.../wireless-technology-10-68253268
Ref. III5: Internet Communications:
https://www.reference.com/.../internet-communication...
Ref. III6. Cryptocurrency Wallet
mailto:https://cryptocurrencyfacts.com/what-is-a-cryptocurrency-wallet/
Ref. III7. Public Key Cryptography (PKC) and Digital Signatures:
https://link.medium.com/VJmb0V4kebb

4) Quantum Secure Voting. The new voting system includes: **a)** voting events, **b)** genesis (i.e., creation) of voting tokens and ballots, **c)** registration, **d)** distribution of tokens and ballots, **e)** voting, **f)** clearing, and **g)** settlement.

a) Voting Events. A voting event will typically have specific voting times and locations, voters, a centralized voting authority for the voting event, a voting authority for each voting district, candidates, and propositions. Voters and voting authorities will be given instructions on how to interact with each other and with QB's ledger and databases. After registering, each voter would receive from their voting authority, email with a unique voter (V)-wallet, two voter authorization tokens, and a unique secure access link to their voting ballot on QB.info's private clearing database.

Each voting authority would be emailed, from the centralized voting authority, instructions on how to create: ballots for their voting district, a unique ballot (B)-wallet, and a unique voting authority (A)-wallet for voter authorization tokens.

Timeless

Ballots would include a Quick Response (QR) code representing the public address of an associated B-wallet. An unmarked voting ballot formatted with candidates and propositions would be uploaded to QB.info's private clearing database for each voting district. The number of ballots in the database would depend on the estimated adult voting population for each voting district's jurisdiction. The first step for a voter, after accessing their ballot, would be to send a voter authorization token from the voter's V-wallet to the ballot's B-wallet. Then the voter can start marking the ballot with choices.

> I.   *Adding a QR code to each voting ballot, representing the public address of a ballot's B-wallet, is the first level in quantum secure voting.*

> II.   *A registered voter sending a voter authorization token from their V-wallet to a ballot's B-wallet to start voting (i.e., marking the ballot), is the second level in quantum secure voting.*

b) Genesis. This section covers the **i)** genesis (i.e., creation) of the voter authorization tokens by QB.com, and **ii)** genesis of unmarked voting ballots by each voting authority.

i) The QB.com public ledger would have a genesis (G)-wallet with voter authorization tokens for each voting event, a cleared (C)-wallet for cleared tokens, and a settled (S)-wallet for settled tokens. The number of tokens in the G-wallet would be more than 4x the adult voter population for the voting event--considering every voting district. Reason for the 4x is that each voter gets two voter authorization tokens-- one to start marking the ballot and a second for settlement; plus one voter authorization token for the voting authority to confirm and clear a voter's ballot, plus one voter authorization token for settlement signaling, as discussed in **4g).** The total number of tokens in the G-wallet would be proprietary--to help identify fraudulent activity.

ii) Ballots would be created by voting authorities for their voting district. Each voting district could have unmarked ballots that are formatted with their district's candidates and propositions.

Timeless

Each unmarked ballot would also have a QR code representing a ballot's B-wallet public address. The unmarked ballot would be uploaded to the QB.info private clearing database. QB.info would then allocate more than enough unmarked ballots in the database for the voters in the voting authority's jurisdiction. The voting authority would then review QB.info's allocation of access links and unmarked ballots for their voting district, adjust if needed, and request the allocation information be emailed to the voting authority if needed.

> ***III.*** *Two voter authorization tokens sent to each registered voter by a voting authority, the first to start marking a ballot and a second for settlement of a ballot, is the third level in quantum secure voting.*

c) Registration. Voters must register to vote with their local voting authority for each voting event. Voting authorities would have an estimate of the total adult voter population (AVPj) in their voting district's jurisdiction (j). To get registered, voters would have a legal right to vote and show proof of life. A proprietary Know Your Voter (KYV) method, similar to Know Your Client (KYC) in finance, would assess the status of the voter. The KYV method depends on the voting event and voting authorities, and would provide a basis for authorization tokens to be sent to a voter. If there is suspected fraud with more registration requests than the adult voter population in a voting district, the voting process would continue but with suspected ballots flagged in QB.info's private clearing database and processed as described in **5)** below.

Voters must also have a valid ID such as a driver's license, address of residence and/or phone number, and email address. Since the registration process is electronic, the voter would also have a recordable IP address. The valid ID and email address are the basis for emailing to the voter a secure access link to voting instructions.

> ***IV.*** *Registering voters, is the fourth level in quantum secure voting.*
>
> ***V.*** *Flagging ballots for voters that have suspicious registration requests, is a fifth level in quantum secure voting.*

d) Distribution. There are three forms of distribution: **i)** traditional email, **ii)** hypertext for website links, and **iii)** a send with transactions recorded on a distributed ledger.

i) The central authority for each voting event would email to each district's voting authority secure access links for instructions and trouble tickets for their district. Each voting authority would then email, to each registered voter in their district, secure access links for trouble tickets and voting instructions. Instructions would including how to get a digital wallet and voter authorization tokens, and the secure access link to their unmarked ballot in QB.info's private clearing database. The voter would email the voting authority when their voting has completed, and receive email when the voting authority confirms their ballot has been validated.

ii) Voters and voting authorities would receive secure access links for their instructions and trouble tickets on QB.site. Voters would receive secure access link for their ballot on QB.info. Voting authorities would have secure access links to QB.info's private clearing database to review ballot allocations and access addresses, and to review and validate completed voting ballots. Voters receive secure access links from their voting authority for their ballot on QB.info's private clearing database.

iii) From the QB.com's G-wallet, voting authorization tokens would be sent to the voting authority's A-wallet. Two voter authorization tokens would be sent from a voting authority's A-wallet to each registered voter's V-wallet. From the voter's V-wallet, one voter authorization token would be sent to ballot's B-wallet to start voting (i.e., ballot marking), and one voter authorization token would be sent to QB.com's S-wallet for settlement. From the voting authority's A-wallet, one voting authorization token would be sent to QB.com's C-wallet to confirm the completed ballot is valid. From the QB.com's G-wallet, one voter authorization token would be sent to the voter's V-wallet to signal the voter that the settled ballot has been copied to the QB.info's public settled database.

> *VI.* *The voting authority sending two voting authorization tokens and emailing a secure access link to an unmarked ballot on QB.info's private clearing database, to each registered voter, is a sixth level in quantum secure voting.*

e) Voting. Each voter has a secure access link to their ballot on QB.info private clearing database. The voter would first send a voter authorization token from the voter's V-wallet to the ballot's B-wallet. Then the voting choices can be marked on the ballot. When marking is complete, QB.info will compute a cryptographic hash of the ballot, and that hash will be stored next to but not on the ballot. The method used to compute a hash depends on the type of PKC selected for the voting event. A hash is a way to summarize a voting ballot with a fixed length character string of letters and numbers. A slight change in a ballot, such as "a" changed to "b" would have an obvious and drastic change in the hash. If voter entries got changed without the voter's authorization, that problem would be addressed as described in **5)** below.

f) Clearing. When ballot marking is complete, the voter would email the voting authority the ballot status as marked but unconfirmed. The voting authority would then inspect the ballot to confirm the ballot was marked and a voting authorization token was entered in the B-wallet. A second hash for the ballot would be added to QB.info's private clearing database when the voting authority inspects the ballot. The hash data is viewable only be the voter. When the voting authority validates the ballot, it sends a voter authorization token from voting authority's A-wallet to the QB.com's C-wallet. An entry in the C-wallet is the first confirmation for a ballot and means that the ballot has been cleared.

> **VII.** *Adding cryptographic hash, viewable only by the voter, after the voter first completes a ballot and each time there is an inspection of the ballot on the QB.info private clearing database, is the seventh level in quantum secure voting.*

> **VIII.** *Changing a ballot status to cleared when a voter authorization token is sent from a voting authority's A-wallet to QB.com's C-wallet, is the eight level in quantum secure voting.*

g) Settlement. Settlement occurs when the ballot recorded in the QB.info's private clearing database is validated by the voting authority and then by the voter. When validated by the voter, a voter authorization token is sent from the voter's V-wallet to QB.com's S-wallet.

Timeless

An entry in the S-wallet is the second confirmation for a marked ballot, and means that the ballot is settled. A copy of the cleared voting data is then entered in QB.info's public settled database, and the QB.com's public ledger for the ballot shows as settled.

To signal the voter that the ballot has been copied to a new database, a voting authorization token from QB.com's G-wallet is sent to the voter's V-wallet. This notifies the voter that the settled data is in a new public database, and the voter can check the QB.info private cleared database for the new secure access link. The private hash and secure access link are viewable only by the voter. The voting authority and others can view the records and summaries on QB.info public settled database, but would not be able to associate a particular record with a voter—and that means the voter's identity is pseudo-anonymous.

IX.   *Settlement occurs when a voting authorization token is sent from a voter's V-wallet to QB.com's S-wallet and a copy of the cleared data is entered in QB.info public settled database, and that is the ninth level in quantum secure voting.*

X.   *QB.com signals the voter with a voter authorization token sent from their G-wallet to the voter's V-wallet, when the voter's ballot is copied to the public settled database, and that is the tenth level in quantum secure voting.*

5) Trouble, Errors and Fraud. The following includes the process for **a)** trouble ticketing, and **b)** filtering suspected errors from fraudulent activity.

a) Trouble Ticketing. Voters can check if their ballots were entered without error or alteration. If not, the voter can access a trouble ticket that would be processed by a QB trouble resolution staff. The resolution process would determine if there is an error or not. If voting data is not in error, no change would be made to any of QB's ledger or databases. If there was an error, the associated ballot data would be flagged.

Timeless

The next step would determine if the error was unintentional and made by the voter, voting authority, QB, or other process. Appropriate resolution steps would be reviewed with the voter and voting authority, to fix an unintentional error and test the upgrade as part of periodic QB maintenance. When a fix is approved, the appropriate QB ledger or database would be updated and the flag would be removed. If it is determined that the error was intentional, that puts it into a category that is addressed next.

b) Filtering Errors from Fraud. Unintentional errors were discussed above. Intentional errors would be treated the same as intentional fraud. There can be two types of intentional fraud—**i)** illegal votes, and **ii)** legal votes that have been changed without the voter's authorization.

i) If intentional fraud is suspected with an illegal voting ballot, the suspect ballot would be flagged in QB.info's private or public database. The fix would be to create a sting operation (XY) to identify the fraudulent activity and participant(s). The fourth level (**IV. Registration**) in quantum secure voting suggests how sting operation XY would proceed. In the registration and KYV process, there could be a crosscheck with public personal data. Personal data that may be available includes social networking activity, and results from search engine and obituary website inquires.

For personal data that is private, if justified a subpoena process could be started to get data such as income, credit score, welfare benefits, and the owner's location and identity for a phone number, email address, or IP address. If fraud is not confirmed, the flag would be removed from the database. If fraud is confirmed, the flag would remain and a copy of the flagged data would be entered in QB.xyz's private database as evidence for subsequent warrants and prosecution.

ii) When a legal ballot has been changed without the voter's authorization, the suspect ballot would be flagged in QB.info's private or public database. The fix would be to isolate where and how the change occurred and create another sting operation (YZ) to identify the fraudulent device(s) and participant(s).

Timeless

The fifth and eight levels (**V**. Hash Processing + **VIII**. Suspicious Data) in quantum secure voting, suggests how sting operation YZ would proceed. Each time the voting ballot was inspected, an ID, time and date would also be recorded for the inspection, along with the cryptographic hash in the QB.info's private clearing database. The first hash would be when the voter first completed the ballot, the second hash when the voting authority inspected the ballot, the third hash when the ballot was validated by voter, and then another hash for any inspection thereafter. If any hash is not the same as the first, then an error or fraudulent activity can be suspected.

If the filtering process determines there was an unintentional error, such as in computing the hash, then that problem would be fixed as a maintenance issue, and the flag removed from the QB.info's private or public database. If the filtering process determines there was intentional fraud, the flag would remain and a copy of the flagged ballot would be entered in QB.xyz's private database. The private database would continue to track the fraudulent activity until the suspected device(s) or participant(s) can be identified. The QB.xyz's private database of fraudulent activity would then be used as evidence in subsequent warrants and prosecution.

6) Data Sharing. Data that can be shared with different groups includes: **a)** public data, **b)** private data, and **c)** a private dashboard.

a) The QB.com public ledger and QB.info public settled database could be made available to local, national and international media, educational institutions, companies and others--after a voting event has completed, and possibly during the event depending on the central voting authority.

In the USA for a particular voting event (e), the public ledger and database could be viewable with numeric and graphic summaries for a voting district jurisdiction (j), county (co), state (s), nation (USA), candidate (ca), and/or proposition (p). A voter would also be able to see how their vote was time sequenced with other votes on the public ledger. Other countries could have their own preferred naming for geographic areas and methods for viewing public voting data.

Timeless

b) Private data would not be viewable by the public. Suspected errors or fraud would be flagged in QB.info's private or public database. If an unintentional error is confirmed, appropriate action would be taken to fix the error and remove the suspected error flag. If fraud is confirmed, a copy of the flagged data would be stored in QB.xyx's private database and used as evidence for subsequent warrants and prosecution.

c) During a voting event, data would be available on QB.site's private dashboard to a QB security group that watch for suspected errors or fraud. Action to review a suspected activity could be triggered by any of the following conditions:

A. There are more voter registration requests than the estimated adult voter population (AVPj), for a voting authority's district jurisdiction (j),

B. A voter authorization token was not sent from a voter's V-wallet to ballot's B-wallet before the ballot marking started in QB.info's private clearing database,

C. Total voter authorization tokens sent from the QB.com's G-wallet to the voting authority's A-wallet(s) are more than 3x the adult voter population (AVPj), for a voting authority's district jurisdiction (j),

D. Total voter authorization tokens in the QB.com's C-wallet are more than 1/3 the tokens sent from the voting authority's A-wallet, for any jurisdiction (j),

E. Total voter authorization tokens in the QB.com's S-wallet are more than 1/3 the tokens sent to a voting authority's A-wallet, for any jurisdiction (j),

F. Total voter authorization tokens sent from the QB.com's G-wallet are more than 4x the total adult voter population (AVPj), for a voting authority's district jurisdiction (j),

G. Total voter authorization tokens in the QB.com's S-wallet are more than the tokens in the C-wallet, for any jurisdiction (j),

H. A cryptograph hash for a ballot in the QB.info's private clearing database is not the same as the first hash, when the ballot was first completed by the voter,

I. Any attempt by other than the voter to view the cryptographic hash data and/or settled ballot access link on QB.info's private clearing database.

QB summaries of public data could be tailored to represent a subset of the voting data for one or more voting districts. For example, for condition C above and county "co", the G-wallet sent tokens would be for all voting district jurisdictions (j) in county (co). If any of the conditions A to I occur that would trigger further action as described in **5)** above.

7) Contact. The author would be interested in consulting with anyone that is in the process of planning, developing, testing, or deploying a modern voting system. The email contact is johnwnuk@gmail.com.

The material in Appendix III has enough detail that it could be turned into a Request For Proposal by a state or federal government agency, or used as a development plan by a company that appreciates the worldwide need for quantum secure voting.

More References

TIMELESS Background
    Earlier draft version is in the *4 M's eBook CHAPTER I*:
linkedin.com/pulse/money-book-john-wnuk
    TIMELESS eBook: (request password at *johnwnuk@gmail.com*)
jmwnuk.wixsite.com/timeless/ebook
    Risks and Disclosure:
jmwnuk.wixsite.com/digitalassets/disclaimer

More on Fiat and Crypto
    Modern Money Mechanics (i.e., fiat money) by the Federal Reserve:
upload.wikimedia.org/… /4a/Modern_Money_Mechanics.pdf
    40 Years Preparing the Soil for Bitcoin
https://www.danheld.com/blog/2019/1/6/planting-bitcoinsoil-34
    Exchange reviews:
steemit.com/cryptocurrency/… /best-…-exchanges-in-2020-get-ready
    Wallet reviews:
thebalance.com/best-bitcoin-wallets-4160642
    Paper Wallet generator:
walletgenerator.net/#

Examples of Websites with Frequently Updated Data
    coinmarketcap.com…
coinmarketcap.com/
    jmwnuk.wixsite.com…
jmwnuk.wixsite.com/digitalassets
    blockchain.com…
blockchain.com/btc/address/1BRFRuhnQTKg6M2bm1wk5vN1KUobgz4hFg
    linkedin.com…
linkedin.com/pulse/part-4-money-fuel-john-wnuk/
    usdebtclock.org…
usdebtclock.org/world-debt-clock.html

Glossary

1G … nG: 1st … nth Generation. Crypto as: 1G Bitcoin, 2G Ethereum, 3G Cardano, 4G IOTA, 5G Nano, Holo; or mobile wireless communications: 1G Analog, 2G TDMA, 3G CDMA, 4G/5G OFDM evolution.
1xTXID: Single-entry Transaction ID accounting. Clearing and settlement have the same transaction ID.
2FA: 2-Factor Authentication.
2xTXID: Double-entry Transaction ID accounting. Clearing and settlement have separate transaction IDs.
4 M's: Mysteries of Modern Money Mechanics.

Abraxas: A mystical god/devil of the occult world representing opposite behaviors such as good/evil, loving/hateful, honest/corrupt, and so on.
ACH: Automated Clearing House.
ADA: A mined cryptocurrency with market value and transactions recorded on the scalable Cardano blockchain.
AI: Augmented Intelligence for a person. Artificial Intelligence for a device, robot, or computer.
Altcoin: Cryptocurrency other than Bitcoin with market value.
AML: Anti-Money Laundering.
ATM: Automated Teller Machine.

Bankruptcy: Chapter 7: Debtor must sell off nonexempt assets to pay creditors; Chapter 11 or Rehabilitation: Debtor can reorganize their debts to try to re-emerge as a healthy entity; Chapter 13: As part of reorganization, debtor must provide substantial pay back to creditors within a fixed time interval.
BCH: Bitcoin Cash. A mined cryptocurrency with market value and transactions recorded on the BCH blockchain.
BIP: Bitcoin Improvement Proposal.
BIP0032: Standard for Hierarchical Deterministic (HD) Wallets, with a hierarchical tree-like structure for private/public key pairs.
BIP0038: Encrypts the private key on a paper wallet with a passphrase for more security if the wallet is lost or stolen.
BIP0068/0112/0113: Used in the smart contract in Bitcoin's Lightning Network.

Bitcoin Wallet: Paired public address and private key for the BTC cryptocurrency.

block: Group of transactions on a blockchain.

Blockchain: Permanent, distributed, digital, public ledger of worldwide Bitcoin transactions. Lower case "blockchain" or DLT in this book is used for the public ledger of Altcoin transactions.

BTC: Bitcoin. A peer-to-peer Electronic Cash System. A mined cryptocurrency with market value and transactions recorded on the BTC Blockchain.

Cardano: A blockchain for the ADA cryptocurrency.

CDMA: Code Division Multiple Access. A 3G mobile wireless communications technology.

CFTC: Commodities Futures Trading Commission.

chain: Links in a blockchain/DLT based on a hash function that allows transactions to trace back to verified sources (or genesis).

clearing: When a crypto transaction is verified and added to an unconfirmed block. Compare with fiat clearing with ACH.

cold storage: Cryptocurrency on media that does not connect to the Internet.

confirmed block: A block in a blockchain or DLT that has not been rejected due to errors of content or format.

Coordicide: Replaces Coordinator as IOTA's distributed ledger technology.

counterfeit: Fraudulent imitation of something of value such as the forging of fiat money.

coupon: Term used by Holo for their ERC-20 cryptocurrency.

COVID-19: Corona Virus Disease 2019.

CPA: Certified Public Accountant. An accounting and tax professional who has met additional certification requirements.

crypto: Short form of cryptocurrency.

cryptocurrency: Money based on cryptographic methods.

cryptography: Methods for secure communications in the presence of unfriendly third parties.

currency/coins: Component of a modern money system that can include fiat currencies, cryptocurrencies or stablecoins.

CYBERCRYPT: Provider of robust cryptography. Developed the Troika hash function for IOTA's DAG architecture and Trinity wallet.

Timeless

D-Wave Systems, Inc: A quantum computing company based in Canada.

DAG: Directed Acyclic Graph. Use of graph theory in mathematics to trace transaction data from source to destination without a repeating cycle.

DAO: Distributed Autonomous Organization. A decentralized organization of computers not controlled by a government or bank.

decree: An official order by a legal authority.

DHT: Distributed Hash Table as used in Holo (HOT).

DIF: Distributed IPC Facility. A single repeating layer in RINA that allows rapid response to changing loads in mobile communications.

digital: A binary sequence of 0's and 1's or "on" and "off" states to represent data in a computer.

digital signature: Method to encode a message with a private key and decode the message with a public key without revealing the private key.

DLT: Distributed Ledger Technology.

DRE: Direct Recording Electronic

DSA: Digital Signature Algorithm

double spend: A flaw in a money system that allows money to be spent more than once, as in counterfeit fiat money.

ECDSA: Elliptic Curve Digital Signature Algorithm. Method to compute a public key from a private key where the reverse is not feasible, and to digitally sign a message with a private key and not reveal the private key.

EdDSA: Edwards-curve Digital Signature Algorithm. Designed to be faster than ECDSA without sacrificing security.

Ed25519: Digital signing using EdDSA encryption with elliptic Curve25519. It is used in Cardano's HD Daedalus Wallet, Nano, and Holo, and Curve25519 was designed for TLS v1.3.

Electoral System: Each state in the U.S. gets electors based on number of representatives in Congress. Electors cast one electoral vote following the general election and there are 538 electoral votes. The candidate that gets more than half (270 or more) wins the election.

Electronic Cash System: Bitcoin. A protocol on the Internet for the transfer of value between peers, with each transaction recorded on a Blockchain (i.e., worldwide public distributed ledger).

elliptic curve: Curve or plot points on a graph based on the equation $y^2 = x^3 + ax + b$. For Bitcoin and Ethereum cryptocurrencies, $a = 0$ and $b = 7$.

Epoch: Method of dividing time in the blockchain in the Cardano Settlement Layer.

ERC-20 token: Cryptocurrency designed for the Ethereum platform.

ETC: Ethereum Classic. A mined cryptocurrency with market value, and with transactions recorded on the ETC blockchain.

ETH: Ethereum. A mined cryptocurrency with market value and transactions recorded on a programmable Ethereum blockchain. Co-founders were Vitalik Buterin, Anthony Di Iorio, Charles Hoskinson, Mihai Alisie, and Amir Chetrit.

exchange(s): A banking, investment or cryptocurrency facility for user accounts and transactions, transaction records, and tax records.

exchange wallet: A wallet used by a cryptocurrency exchange, such as Coinbase, to add more security to a user's digital wallet.

Facebook: An online social network where people create profiles, share information, and respond to information posted by others.

FDIC: Federal Deposit Insurance Corporation.

Fed: Federal Reserve System. The central bank for the U.S.

Fee-less transaction: Method of transaction verification that does not require miners or mining fees and is similar to free https web browsing.

FFM: Fast, Fee-less, and Miner-less transaction verification.

fiat: Money with the value based on a government decree and not based on a tangible commodity such as gold.

FICO® Score: Credit rating by Fair Isaac Corporation. Score <580 is poor, 580-799 is fair to very good, and 800+ is exceptional.

FinCEN: Financial Crimes Enforcement Network.

fintech: financial technology.

forks: Planned enhancements or maintenance points in a cryptocurrency development roadmap.

Fractional Reserve Lending: Bank deposits are lent to others to generate interest income and are not stored in a bank vault.

GBP: British pound sterling.

genesis: The source or origin of something of value. The first block in a blockchain.

gold standard: Monetary system where a country's paper money can be converted to a fixed amount of gold.

halving: A point in time when a block reward is cut in half.

hard fork: Change in a blockchain or DLT to include a secondary chain.

hash: Cryptographic method to compute a fixed length text from a message of any length, and used to identify if a message has been altered.
HD Wallet: Hierarchical Deterministic Wallet. A system for deriving private-public key pairs from a starting seed as defined in BIP0032.
heyday: Time or phase of greatest prosperity, success, or vigor.
HHO: Water as gas includes two parts Hydrogen and one part Oxygen.
HOT: Holo: A pre-mined cryptocurrency with market value. Uses double-entry Transaction ID accounting with debit and credit transactions.
HTLC: Hashed Time Lock Contract. A proposal for Bitcoin's Lightning Network to improve scalability.
HTTP: HyperText Transfer Protocol for non-secure Internet browsing.
HTTPS: HTTP for secure Internet browsing with Secure Socket Layer (SSL) or Transport Layer Security (TSL).

IBM: International Business Machines.
ID: Identification.
IEEE 802.11: Institute of Electronic and Electrical Engineers standard 802.11. Part of the IEEE 802 set of Local Area Network (LAN) protocols for Wi-Fi computer communications.
immutable: Not changing or unable to be changed.
Internet: Worldwide network of computers and standard protocols to route packets of data and provide access to information.
IOT: Internet Of Things.
IOTA: A pre-mined cryptocurrency with market value. The protocol uses single-entry Transaction ID (1xTXID) accounting on the scalable Tangle DLT. Co-founders were David Sonstebo and Dominik Schiener.
IOTA Coordicide: Part of IOTA's plan for scalability with an outline at *coordicide.iota.org/*.
IOTA Tangle DLT Main Net: IOTA transaction model at *tangle.glumb.de*.
IP: Internet Protocol or Intellectual Property.
IPC: Inter-Processor Communications as in DIF for RINA.
IRS: Internal Revenue Service.
ISP: Internet Service Provider.

JPM Coin: J. P. Morgan stablecoin.

KDP: Kindle Direct Publishing.
Keccak-256: Hash used by Ethereum that uses Sponge construction.
KYC: Know Your Client or Know Your Customer.

LAN: Local Area Network.

law of networks: n(n-1)/2: Started with phone networks with n equal to the number of phones. With n = 1 phone, value is 0. With large n, value gets exponentially large with value near (n^2)/2.

Libra: Facebook's plan for a new stablecoin and blockchain that provides a platform for financial services innovation.

Lightning Network: A layer in the Bitcoin protocol that enables fast off-chain transactions between participating nodes.

MAC: Media Access Control. Used for physical device addresses. A MAC address is a number assigned by the manufacturer vs. an IP address that is a number assigned to a connection in a network.

markets: Pairs for trading transactions including: fiat/fiat (i.e. USD/EUR), fiat/crypto (i.e. USD/BTC), and crypto/crypto (i.e. BTC/ETH).

message: Data in a crypto transaction that includes at least a from public wallet address, to public wallet address, and the amount to be transferred between wallets.

mining: Method to verify transactions (i.e., clear for entry to a blockchain and settle in a confirmed block), and to produce new crypto by solving a math problem.

mining fees: The payment to miners to verify transactions.

Miner-less: Method for transaction verification with an agent or client that does not require miners or mining fees.

MIT: Massachusetts Institute of Technology.

Money: Defined as *three services*: medium of exchange, store of value, unit of account; and *six characteristics*: durable, portable, divisible, uniform, limited supply, and acceptable.

Money System: Defined as six components: currency/coins, wallets, reserve systems, exchanges, markets, and transactions.

MTL: Money Transmitter License.

NANO: Nano: A pre-mined cryptocurrency with market value. Nano uses a DAG-based Block Lattice with double-entry Transaction ID accounting to include a signed send transaction and a signed receive transaction.

national financial system: A Central Bank for a nation.

NE: North East

Novi: Digital wallet for Facebook's Libra cryptocurrency.

NSA: National Security Agency.

OFDM: Orthogonal Frequency Division Multiplexing. A technique for wireless communications used in 4G, 5G and future 6G mobile wireless communications.

Off-chain: Transactions are not recorded on a blockchain or DLT.

On-chain: Transactions are recorded on a blockchain or DLT.

Owner of value in a wallet: Whoever has the private key.

PBOC: People's Bank Of China.

peer-to-peer: Interchange between like things as in person-to-person or Bitcoin wallet-to-Bitcoin wallet.

PKC: Public Key Cryptography. Method to encode a digital signature with a private key but not reveal the private key, and to decode a digital signature with a public key to verify a message is valid.

POS: Proof Of Stake. An alternative to POW.

POW: Proof Of Work. Computation performed by miners before new cryptocurrencies are produced.

pre-mined: A term used for crypto that is not mined and typically made available in the genesis block.

private key: Random character string to compute a public key and public address, or digitally sign a message.

PSA: Payment Services Act (in Japan).

pseudo-anonymous: Cryptocurrency wallet transactions are traceable on a blockchain or DLT, but are not traceable to the owner of the wallet.

public address: For Bitcoin, a RIPEMD-160 and SHA-256 hash of a public key. To avoid visual errors, a Bitcoin public address does not have number zero "0," upper case for o "O," lower case for L "l," or upper case for i "I."

public key: Computed from a private key using a variety of methods including ECDSA and Ed25519.

QR: Quick Response.

quantum computer: A computer with data represented as qubits, and theoretically superior to computers with data represented as bits.

quantum security: Protocol design that makes it computationally unfeasible for a quantum computer to reveal a wallet's private key.

qubit: a quantum bit. Counterpart to a bit in classical computing. A qubit is the basic unit of information in a quantum computer.

Quorum: DLT for the JPM Coin.

R&D: Research & Development.
RACE: Research on Advanced Communications in Europe.
RINA: Recursive Inter-Network Architecture. A common repeating function for scalability. It was developed for mobile wireless communications and is used in the Cardano ADA cryptocurrency.
RIPEMD-160: RACE Integrity Primitives Evaluation Message Digest-160. Hash of SHA-256 hashed public key to create a public address.
Ripple: Blockchain for the XRP pre-mined cryptocurrency.
RMC: Reserve Market Cryptocurrency.
RSC: Reserve Stable Cryptocurrency.
RTGS: Real Time Gross Settlements. Method of fiat money transfer between banks. There is a separate RTGS for each sovereign nation.

SARS-CoV-2 : Severe Acute Respiratory Syndrome Corona Virus 2.
Sat: Satoshi: Smallest unit of Bitcoin with 1 Sat = 0.00000001 BTC and 1 BTC = 100,000,000 Sat.
scalable: Designs that can increase capacity with increased transaction load. For example, a design that can increase from 10 transactions per second (tps) to 100 tps or 1,000 tps, as needed.
SDW: Smart Digital Wallet
SEC: Securities and Exchange Commission.
seigniorage: Difference between face value and production cost of fiat money.
secp256k1: Elliptic curve parameters associated with Koblitz curve $y^2 = x^3 + 7$ used in Bitcoin and Ethereum.
SegWit: Segregated Witness
settlement: An unconfirmed block of transactions becomes confirmed on a blockchain. Compare with RTGS for fiat settlements between banks.

SHA-256: Secure Hash Algorithm-256. Cryptographic hash that generates a 256-bit (32-byte) signature for a message of arbitrary length.
Sharding: A database partitioning technique to improve scalability.
signing algorithm: Use of public and private keys for secure message transmission over an insecure network.
Smart Contract: A digital protocol to enforce the performance of a contract. It allows transactions to occur without third parties.
soft fork: An agreed to change in a crypto protocol that does not result in a secondary blockchain or DLT.

sovereignty: Authority of a nation to shape its own destiny.

Sponge hash: A function that can take an input bit stream of any length and produce an output bit signature of any desired length.

SRS: Sovereign Reserve System: As in a generic form of a Federal Reserve System to produce, store, distribute, secure money or something of value.

SRS-C: Sovereign Reserve System-Crypto: For RSC and RMC.

SRS-F: Sovereign Reserve System-Fiat: Dollar in the U.S., Euro in Europe, …

SSL: Secure Socket Layer: Encrypted link between server and browser.

Stable Cryptocurrency: stablecoin: Cryptocurrency with value pegged to fiat or other tangible source to remove volatility.

Sustainable: The ability to exist or endure considering a variety of controllable and uncontrollable conditions.

sweep: Computer or smartphone optically scanning a private key to withdraw value stored in the associated digital wallet.

TBD: To Be Determined.

TDMA: Time Division Multiple Access. A method for mobile wireless communications, typically referred to as 2G.

Tether: See USDT.

TLS: Transport Layer Security. Naming of SSL after SSL 3.0.

ticker symbol: Sequence of characters representing traded assets.

token: Cryptocurrency based on the ERC-20 Ethereum platform.

tps: transactions per second.

transaction(s): Transfer of ownership of value. Data in a transaction includes a from (or debit) wallet address, to (or credit) wallet address, public key, message signed with a private key and to be verified with public key. Transactions are recorded on a distributed ledger along with the exchange, computer or smartphone where the transaction occurred.

Trinity Wallet: The IOTA wallet for smartphones and computers.

Troika: A cryptographic hash function for the use in IOTA and designed by CYBERCRYPT.

UK: The United Kingdom.

unconfirmed block: The newest block in a blockchain or DLT with cleared but unconfirmed (i.e., not settled) transactions.

UPS: United Parcel Service.

U.S.: United States.

USA: United States of America.

USD: United States Dollar.
USDC: USD Coin. A stablecoin or pre-mined cryptocurrency with stable value developed by Circle and Coinbase in the CENTRE consortium.
USDT: Tether. A stablecoin or pre-mined cryptocurrency with stable value issued by Tether Limited.
UTXO: Unspent Transaction Output. Can be spent as an input in a new transaction.

wallet(s): fiat wallet/purse or crypto digital wallet: A fiat wallet (or purse) typically contains fiat cash and coins, one or more IDs, and credit and debit cards. A crypto digital wallet is a paired public address and private key. Digital wallets can be on an exchange with the exchange owning the private key; or on a computer, smartphone, or paper that includes the public address and private key.
WARS: Wuhan Acute Respiratory Syndrome.
whale: Person or group with enough resources to affect the price of Bitcoin or Altcoin by buying or selling large quantities.
Wi-Fi: An Alliance of over 500 members, with wireless local area networking based on the IEEE 802.11 family of standards.
Winternitz: An algorithm that allows speed and storage tradeoffs to meet security objectives in digital signatures. Used by IOTA.
WIP: Work In Progress.
www: WWW: World Wide Web: Network-accessible information with links defined with http://www or secure https://www .

XRP: Pre-mined cryptocurrency with market value and transactions recorded on the Ripple blockchain.

Epilogue

Paper Wallets and Quick Response Codes.

Digital wallets can be electronic or on paper. Bitcoin paper wallets can be created at *Bitaddress.org*. A wallet has a paired public address and private key with corresponding QR codes. The private key can be BIP0038 passphrase protected for more security. Paper wallets can be used for gifts, secure "cold" storage, and education on the crypto process by observing wallet transactions on the Blockchain ledger. Two examples show how QR codes are used with a Bitcoin app on a computer or smartphone.

The first example is a Bitcoin paper wallet with passphrase protection as shown in Figure 2.

Figure 2. Paper Wallet with Passphrase Protection

The public address is for deposits via "send" to the QR code corresponding to: *1GX81UyD8EYwgZg3xFE3djLjiLAzMJSvka*. The QR code on the right is the passphrase protected private key corresponding to:

6PnNFmxkpxzLc6gHuw2NYLNfyfDhrpBAmcdUVWsdcnktoJSpNg7ztqzAae.

To withdraw value, "sweep" the private key with a Bitcoin app, then enter the passphrase (not shown but will be provided by first requesting non-family person with email to *johnwnuk@gmail.com*). Deposit and withdrawal transactions are recorded on the Blockchain at *(Ref. Ep1).*

Ref. Ep1:
blockchain.com/btc/address/1GX81UyD8EYwgZg3xFE3djLjiLAzMJSvka

Epilogue

The second example is a Bitcoin paper wallet without passphrase protection as shown in Figure 3.

Figure 3. Paper Wallet without Passphrase Protection

To deposit value, "send" to the wallet's public address QR code on the left corresponding to: *1BRFRuhnQTKg6M2bm1wk5vN1KUobqz4hFq*.

The private key QR code for withdrawals on the right corresponds to: *L15fJvHwywpb4n5kssJr4BioA2ozmN4djsn2BZBV6pB28QafAgsJ*

To withdraw value, sweep the private key with a Bitcoin app. Deposit and withdraw transactions are recorded on the Blockchain at (*Ref. Ep2*).

Ref. Ep2:
blockchain.com/btc/address/1BRFRuhnQTKg6M2bm1wk5vN1KUobqz4hFq

For the prevention of double spends, if there were two or more copies of the wallet in Figure 3, the first copy used to withdraw value would leave an empty wallet as recorded on the Blockchain. If a second attempt was made to withdraw value (i.e., a double spend), the wallet in Figure 3 would be empty and the transaction would not be allowed.

Knowledge of how wallets work is key to understanding crypto and blockchains. Making deposits and withdrawals and observing those transactions on a blockchain, could be the Wow factor that provides the motivation to learn more.

The Coiner Poem

Should I buy or should I sell
Every day, a new story to tell.
Bitcoin (BTC) or Ethereum (ETH) will likely win the day
Always with more crypto (currencies) in the fray.

A Satoshi (Sat) is a basic Bitcoin unit and 100 million Sat = 1 BTC
And when 1 Sat = $0.01 then 1 BTC = $1 million, golly gee.
A Blockchain is a distributed public ledger
That tracks wallet transactions and is very clever.

A wallet is a paired public address and private key
Like a mailbox address and owner's key, you see.
A wallet in a computer or smartphone is not tangible to hold
But is more like a bank that can store digital gold.

There are lots of hackers, since the Internet is not secure
But modern cryptography is there as a cure --
With a digital signature for every wallet transaction
Along with a public key to verify each signing action.

For those that crave stability, volatility is feared
Caused by crypto whale trading and people that are weird.
If you do play, security and 2FA are a priority
Otherwise you risk losing ownership authority.

Epilogue

For financial diversity, balance fiat for the future
With wallets in cold storage plus crypto on a computer.
Fiat has synergy with crypto--the new money
But most have no idea or just think it's funny.

Fiat value is based on sovereign dictation
That debt justifies money creation.
In contrast, crypto value is based on math
With an elliptic curve providing a new path.

The Internet transfers packets without a central controller
Crypto transfers are similar without a 3rd party broker.
For those that are sad because they did not anticipate
Take time to learn more and "be ready" to participate.

Acknowledgement:
The Kipling Society, *The Coiner*, Circa 1611

Best wishes for health and happiness

John grew up in Pittston, a small town in NE Pennsylvania, with his four sisters. His favorite high school stories are about his Wilson Cloud Chamber science project and football legend "Jumpin" Johnny McHale. In college, John studied engineering and has degrees from The Pennsylvania State University and Case Western Reserve University. After college, John worked as a Systems Engineer in communications technology with AT&T and Motorola.

He is part of the American dream. John's dad immigrated to the U.S. in the early 1900's. While his parents were only able to get elementary educations, John was able to get a graduate degree and contribute to the communications industry. Now he has become an author that likes to unravel the evolution towards sustainable economies.

John has two daughters and four grandchildren. After retirement, John teamed with his oldest grandson on a dream to build a dune buggy powered with water. That dream included experimenting with electrolysis--a method of converting water to HHO gas. After an explosive test with the HHO gas, that dream continues, but has been put on hold for safety issues.

In his spare time, John enjoys being an author, inventor, and website developer. Registered copyrights include those in this book, and © 2018, TXu 2-125-326: *Leveraging Time*. Patent US10204378B1 on *Flexible Payment Services* was granted in 2019. His website *jmwnuk.wixsite.com/digitalassets* contains background material for this book.

John's wife Sherri contributed to this book through her review as a professional editor. John and Sherri enjoy time with family and friends, and love to travel. The next trip is always part of their travel planning.

Past
- Barter
- Evolution

Present
- Crypto Value
- Fiat Debt

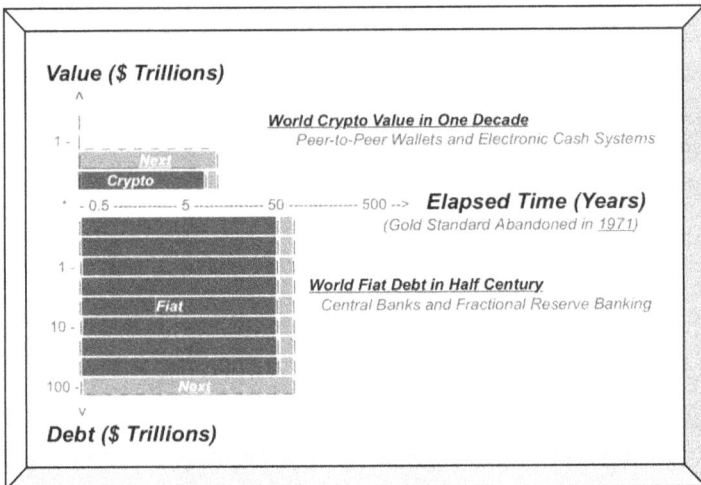

Future
- Verifiable Transactions
- Sustainable Economies

www.ingramcontent.com/pod-product-compliance
Lightning Source LLC
Chambersburg PA
CBHW031329040426

42443CB00005B/267